SOLUTIONS MANUAL
TO ACCOMPANY
LINEAR ALGEBRA

SOLUTIONS MANUAL TO ACCOMPANY LINEAR ALGEBRA
Ideas and Applications

Fourth Edition

RICHARD PENNEY
Purdue University

Published by John Wiley & Sons, Inc., Hoboken, New Jersey.
Published simultaneously in Canada.

For general information on our other products and services please contact our Customer Care
Department with the U.S. at 877-762-2974, outside the U.S. at 317-572-3993 or fax 317-572-4002.

Wiley also publishes its books in a variety of electronic formats. Some content that appears in print,
however, may not be available in electronic format.

Library of Congress Cataloging-in-Publication Data:

Penney, Richard C.
   Linear algebra : ideas and applications / Richard Penney. – Fourth edition.
      pages cm
   Includes index.
   ISBN 978-1-118-90958-4 (cloth)
 1. Algebras, Linear–Textbooks.   I. Title.
   QA184.2.P46 2015
   512'.5–dc23

                                                      2015016650

Printed in the United States of America

10 9 8 7 6 5 4 3 2 1

CONTENTS

STUDENT MANUAL   1

1   SYSTEMS OF LINEAR EQUATIONS   3

   1.1   The Vector Space of $m \times n$ Matrices / 3
         1.1.2   Applications to Graph Theory I / 7
   1.2   Systems / 8
         1.2.2   Applications to Circuit Theory / 11
   1.3   Gaussian Elimination / 13
         1.3.2   Applications to Traffic Flow / 18
   1.4   Column Space and Nullspace / 19

2   LINEAR INDEPENDENCE AND DIMENSION   26

   2.1   The Test for Linear Independence / 26
   2.2   Dimension / 33
         2.2.2   Applications to Differential Equations / 37
   2.3   Row Space and the Rank-Nullity Theorem / 38

3 LINEAR TRANSFORMATIONS 43

3.1 The Linearity Properties / 43

3.2 Matrix Multiplication (Composition) / 49

   3.2.2 Applications to Graph Theory II / 55

3.3 Inverses / 55

   3.3.2 Applications to Economics / 60

3.4 The *LU* Factorization / 61

3.5 The Matrix of a Linear Transformation / 62

4 DETERMINANTS 67

4.1 Definition of the Determinant / 67

4.2 Reduction and Determinants / 69

   4.2.1 Volume / 72

4.3 A Formula for Inverses / 74

5 EIGENVECTORS AND EIGENVALUES 76

5.1 Eigenvectors / 76

   5.1.2 Application to Markov Processes / 79

5.2 Diagonalization / 80

   5.2.1 Application to Systems of Differential Equations / 82

5.3 Complex Eigenvectors / 83

6 ORTHOGONALITY 85

6.1 The Scalar Product in $\mathbb{R}^n$ / 85

6.2 Projections: The Gram-Schmidt Process / 87

6.3 Fourier Series: Scalar Product Spaces / 89

6.4 Orthogonal Matrices / 92

6.5 Least Squares / 93

6.6 Quadratic Forms: Orthogonal Diagonalization / 94

6.7 The Singular Value Decomposition (SVD) / 97

6.8 Hermitian Symmetric and Unitary Matrices / 98

7 GENERALIZED EIGENVECTORS   **100**

7.1   Generalized Eigenvectors / 100

7.2   Chain Bases / 104

8 NUMERICAL TECHNIQUES   **107**

8.1   Condition Number / 107

8.2   Computing Eigenvalues / 108

STUDENT MANUAL

CHAPTER 1

SYSTEMS OF LINEAR EQUATIONS

1.1 THE VECTOR SPACE OF $m \times n$ MATRICES

Problems begin on page **17**

EXERCISES

1.1

(a) $\begin{bmatrix} -1 & -4 & -7 & -10 \\ 1 & -2 & -5 & -8 \\ 3 & 0 & -3 & -6 \end{bmatrix}$, $[3, 0, -3, -6]$, $\begin{bmatrix} -4 \\ -2 \\ 0 \end{bmatrix}$

(b) $\begin{bmatrix} 1 & 8 \\ 4 & 32 \\ 9 & 72 \end{bmatrix}$, $[9, 72]$, $\begin{bmatrix} 8 \\ 32 \\ 72 \end{bmatrix}$

(c) $\begin{bmatrix} \frac{1}{2} & -\frac{1}{2} \\ -\frac{1}{2} & -\frac{1}{2} \\ -1 & 0 \end{bmatrix}$, $[-1, 0]$, $\begin{bmatrix} -\frac{1}{2} \\ -\frac{1}{2} \\ 0 \end{bmatrix}$

Solutions Manual to Accompany Linear Algebra: Ideas and Applications, Fourth Edition. Richard Penney.
© 2016 John Wiley & Sons, Inc. Published 2016 by John Wiley & Sons, Inc.

1.3 $C = A + B$.

1.5 **(a)** $[1, 1, 4] = [1, 1, 2] + 2[0, 0, 1]$

(c) $\begin{bmatrix} 1 & 2 \\ 0 & 0 \end{bmatrix} = 0 \begin{bmatrix} 0 & 0 \\ 1 & 0 \end{bmatrix} + \begin{bmatrix} 1 & 0 \\ 0 & 0 \end{bmatrix} + 2 \begin{bmatrix} 0 & 1 \\ 0 & 0 \end{bmatrix}$

(d) $\begin{bmatrix} 1 \\ 2 \\ 3 \end{bmatrix} = \begin{bmatrix} 4 \\ 5 \\ 6 \end{bmatrix} - \begin{bmatrix} 3 \\ 3 \\ 3 \end{bmatrix} + 0 \begin{bmatrix} 9 \\ 12 \\ 15 \end{bmatrix}$

(f) $-3 \begin{bmatrix} 3 & -1 & 2 \\ 0 & 1 & 4 \end{bmatrix} = \begin{bmatrix} -9 & 3 & -6 \\ 0 & -3 & -12 \end{bmatrix}$

1.6 $P_2 = P_5 - P_1 - P_3 - P_4$, where P_i is the ith row of P.

1.9 Each vector has a nonzero entry in the positions where the other two vectors have zeros.

1.10 Suppose first that

$$A_3 = xA_1 + yA_2$$

Then

$$[0, 0, 8] = [x, 2x + 5y, 3x + 6y]$$

Equating entries yields the system

$$0 = x$$
$$0 = 2x + 5y$$
$$8 = 3x + 6y$$

From the first equation, $x = 0$. It then follows from the second equation that $y = 0$, which is impossible due to the third equation. Thus the independence is proved.

1.12 $[1, -1, 0], [1, 0, 0], [2, -2, 0]$, and $[4, -1, 0]$ all belong to the span. $[1, 1, 1]$ does not because its last entry is nonzero.

1.13 **(a)** $-2X + Y = [1, 1, 4]$ (other answers are possible)].

(b) Let $[x, y, z] = aX + bY = [-a - b, a + 3b, -a + 2b]$ and substitute into $5x + 3y - 2z$. You should get 0.

(c) Any point $[x, y, z]$ that does not solve the equation $5x + 3y - 2z = 0$ will work—for example, $[1, 1, 1]$.

1.16 No. From the second and third entries $aX + bY$ has positive entries only if both a and b are negative; hence the first entry is negative.

1.19 **(c)** No. If $f(x) = a \cos x + b \sin x$, then $f(0) = a$ and $f(\pi) = -a$ would both be positive which is impossible.

1.20 **(c)** Since the intersection of two planes through the origin is a line, the span of $\{X, Y\}$ must be a line. Hence let $X = [x, y, x]$ where $z \neq 0$, any $Y = cX$ where $c \neq 1$.

1.23 For the first part, use various values of a, b, and c in $aX + bY + cZ$. For the second part note that for all scalars $a, b,$ and c the $(2, 1)$ entry of $aX + bY + cZ$ is zero. Hence any matrix W in $M(2, 2)$ such that $W_{2,1} \neq 0$ will not be in the span.

1.25 Let V and W be elements of the span. Then $V = aX + bY$ and $W = cX + dY$. Then for $s, t \in \mathbb{R}$, $sV + tW = (as + ct)X + (bs + dt)Y$ which belongs to the span of X and Y.

1.26 Let the columns of A be A_i, $i = 1, 2, 3$. Then $3A_3 - A_2 = A_1$.

1.29 Let one row be a linear combination of the other rows. This is easily done keeping all entries non-zero.

1.31 **(a)** Yes: $D = 5A - 2B$.

   (b) Yes: $B = A - C$, so $D = 3A + 2C$.

   (c) You can say nothing about the dependence of A and B. Given A and B, dependent or not, let $C = A - B$ and $D = 2A + B + 3C$.

1.33 **(b)** $\sinh x = \frac{1}{2}(e^x - e^{-x}) = \frac{1}{4}(2e^x) - \frac{1}{6}(3e^{-x})$

   (d) From the double angle formula for the cosine function $\cos(2x) = -\sin^2 x + \cos^2 x$.

   (f) $(x + 3)^2 = x^2 + 6x + 9$.

   (h) From the angle addition formulas for the sine and cosine functions

$$\sin\left(x + \frac{\pi}{4}\right) = \sin(\pi/4)\cos x + \cos(\pi/4)\sin x$$

$$= \frac{\sqrt{2}}{2}(\cos x + \sin x)$$

$$\cos\left(x + \frac{\pi}{4}\right) = \cos(\pi/4)\cos x - \sin(\pi/4)\cos x$$

$$= \frac{\sqrt{2}}{2}(\cos x - \sin x)$$

$$\sin x = \frac{1}{\sqrt{2}}\left(\sin\left(x + \frac{\pi}{4}\right) - \cos\left(x + \frac{\pi}{4}\right)\right)$$

(i)

$$\ln[(x^2 + 1)^3/(x^4 + 7)] = 3\ln(x^2 + 1) - \ln(x^4 + 7)$$
$$= 3\ln(x^2 + 1) - 2\ln\sqrt{x^4 + 7}.$$

1.34   The span is the set of polynomials of degree $d \leq 2$. Any pair of such polynomials answers the first question.

1.36   **(a)**   Let $B = \begin{bmatrix} x & y \\ z & w \end{bmatrix}$. Then

$$A + B = \begin{bmatrix} x + a & y + b \\ z + c & w + d \end{bmatrix} = \begin{bmatrix} x & y \\ z & w \end{bmatrix}$$

Hence, $x + a = x$, $y + b = y$, $z + c = c$, and $w + d = d$, which imply that $x = y = z = w = 0$. Hence, $B = 0$.

(b)   Solved similarly to (a).

1.37   See Example 1.4 on page 12 of the text. For example to prove (i) we let $X \in M(n, m)$, $X = [x_{ij}]$. For scalars k and l

$$(k + l)X = [(k + l)x_{ij}]$$
$$= [kx_{ij} + lx_{ij}]$$
$$= [kx_{ij}] + [lx_{ij}] = kX + lX$$

1.40   The steps are as shown below. The vector space properties used were:

Step 1 (a) and (e),

Step 2 (c) and (e),

Step 3 (b), (e), and Proposition 1.2 on page 15,

Step 4 (b), (e), and (g),

Step 5 (f),

Step 6 (h), (g),

Step 7 (j).

$$-(aX) + (aX + (bY + cZ)) = -(aX) + 0$$
$$(-(aX) + aX) + (bY + cZ) = -(aX)$$
$$0 + (bY + cZ) = -1(aX)$$

$$bY + cZ = (-a)X$$
$$(-a)^{-1}(bY + cZ) = (-a)^{-1}((-a)X)$$
$$(((-a)^{-1}b)Y + ((-a)^{-1}c)Z) = 1X$$
$$\left(-\frac{b}{a}\right)Y + \left(-\frac{c}{a}\right)Z = X$$

1.1.2 Applications to Graph Theory I

Problems begin on page **27**

SELF-STUDY QUESTIONS

1.1 The matrices for parts (a), (b), and (c) are respectively

$$
\begin{bmatrix} 0 & 2 & 0 & 1 \\ 1 & 0 & 0 & 1 \\ 0 & 1 & 0 & 0 \\ 0 & 0 & 1 & 0 \end{bmatrix},
\begin{bmatrix} 0 & 1 & 0 & 1 & 2 \\ 1 & 0 & 1 & 0 & 1 \\ 0 & 1 & 0 & 1 & 0 \\ 0 & 0 & 1 & 0 & 0 \\ 0 & 0 & 0 & 1 & 0 \end{bmatrix},
\begin{bmatrix} 0 & 1 & 0 & 0 & 0 & 0 & 0 \\ 0 & 0 & 1 & 0 & 0 & 0 & 0 \\ 0 & 0 & 0 & 1 & 0 & 0 & 0 \\ 0 & 0 & 0 & 0 & 1 & 0 & 0 \\ 0 & 0 & 0 & 0 & 0 & 1 & 0 \\ 0 & 0 & 0 & 0 & 0 & 0 & 1 \\ 1 & 1 & 1 & 1 & 1 & 1 & 0 \end{bmatrix}
$$

1.2 Possible routes are as in Figure 1.1

1.3 An airline would not have a flight from a given city A to itself.

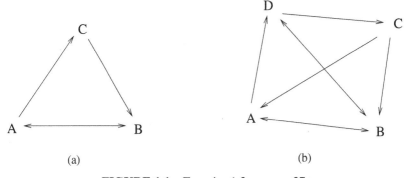

    (a)   (b)

FIGURE 1.1 Exercise 1.2 on page 27.

EXERCISES

1.45 We list the vertices in the order MGM, MGF, PGM, PGF,F, M, S1, S2, D1, D2.

$$\begin{bmatrix} 0 & 1 & 0 & 0 & 0 & 1 & 0 & 0 & 0 & 0 \\ 0 & 0 & 0 & 0 & 0 & 0 & 0 & 0 & 0 & 0 \\ 0 & 0 & 0 & 1 & 1 & 0 & 0 & 0 & 0 & 0 \\ 0 & 0 & 0 & 0 & 1 & 0 & 0 & 0 & 0 & 0 \\ 0 & 0 & 0 & 0 & 0 & 0 & 1 & 1 & 0 & 0 \\ 0 & 0 & 0 & 0 & 1 & 0 & 1 & 1 & 1 & 1 \\ 0 & 0 & 0 & 0 & 0 & 0 & 0 & 1 & 1 & 1 \\ 0 & 0 & 0 & 0 & 0 & 0 & 0 & 0 & 1 & 1 \\ 0 & 0 & 0 & 0 & 0 & 0 & 0 & 0 & 0 & 0 \\ 0 & 0 & 0 & 0 & 0 & 0 & 0 & 0 & 0 & 0 \end{bmatrix}$$

1.46
$$\begin{bmatrix} 2 & 0 & 2 & 1 \\ 1 & 1 & 2 & 1 \\ 2 & 0 & 2 & 1 \\ 0 & 3 & 0 & 3 \end{bmatrix}$$

1.2 SYSTEMS

Problems begin on page **38**

In our solutions, Roman numerals refer to equations in a system. Thus, for example "IV" is the fourth equation in a given system.

EXERCISES

1.49 X is a solution since $4 \cdot 1 - 2 \cdot 1 - 1 - 1 = 0$ and $1 + 3 \cdot 1 - 2 \cdot 1 - 2 \cdot 1 = 0$. Y is not since $1 + 3 \cdot 2 - 2 \cdot (-1) - 2 \cdot 1 = 7$.

1.50 Let $Z = aX + bY = [a + b, a + 2b, a - b, a + b]$. This solves the system if and only if

$$4(a + b) - 2(a + 2) - (a - b) - (a + b) = 0$$
$$(a + b) + 3(a + 2) - 2(a - b) - 2(a + b) = 0$$

which simplifies to

$$0 = 0$$
$$7b = 0$$

Hence Z is a solution to the system if and only if $b = 0$.

1.52 Let

$$Z = aU + bV = [ax + bx', ay + by', az + bz', aw + bw']^t$$

Substituting Z into the system yields

$$4(ax + bx') - 2(ay + by') - (az + bz') - (aw + bw') = 0$$
$$(ax + bx') + 3(ay + by') - 2(az + bz') - 2(aw + bw') = 0$$

which simplifies to

$$a(4x - 2y - z - w) + b(4x' - 2y' - z' - w') = 0$$
$$a(x + 3y - 2z - 2w) + b(x' + 3y' - 2z' - 2w') = 0 \qquad (1.1)$$

Since both U and V are solutions,

$$4x - 2y - z - w = 0$$
$$x + 3y - 2z - 2w = 0$$

and

$$4x' -' 2y' -' z' -' w = 0$$
$$x' +' 3y' -' 2z' -' 2w = 0$$

Hence each term in (1.1) is zero showing that Z is a solution to the system for all a and b.

1.54 Let

$$Z = aU + bV = [ax + bx', ay + by', az + bz', aw + bw']^t$$

Substituting Z into the system yields

$$4(ax + bx') - 2(ay + by') - (az + bz') - (aw + bw') = 1$$
$$(ax + bx') + 3(ay + by') - 2(az + bz') - 2(aw + bw') = 2$$

which simplifies to

$$a(4x - 2y - z - w) + b(4x' - 2y' - z' - w') = 1$$
$$a(x + 3y - 2z - 2w) + b(x' + 3y' - 2z' - 2w') = 2 \tag{1.2}$$

Since both U and V are solutions,

$$4x - 2y - \ z - \ w = 1$$
$$x + 3y - 2z - 2w = 2$$

and

$$4x' -' 2y' -' \ z' -' \ w = 1$$
$$x' +' 3y' -' 2z' -' 2w = 2$$

Hence (1.2) is equivalent with

$$a + b = 1$$
$$2a + 2b = 2$$

Hence Z is a solution if and only if $a + b = 1$.

1.55 In each exercise we give the reduced echelon form of the coefficient matrix followed by the translation vector and the spanning vectors. Where relevant, we list the dependency relations among the rows where the rows are denoted I, II, ...

(b) $\begin{bmatrix} 1 & 0 & 0 & -59/9 \\ 0 & 1 & 0 & 20/9 \\ 0 & 0 & 1 & 8/9 \end{bmatrix}$, $\frac{1}{9}[-59, 20, 8]^t$, 0.

(c) $\begin{bmatrix} 1 & 0 & 17/2 & 1 \\ 0 & 1 & -5/2 & 0 \\ 0 & 0 & 0 & 0 \end{bmatrix}$, $[1, 0, 0]^t$, $\frac{1}{2}[-17, 5, 2]^t$, 2I + II = III.

(d) $\begin{bmatrix} 1 & 0 & 17/2 & 0 \\ 0 & 1 & -5/2 & 0 \\ 0 & 0 & 0 & 1 \end{bmatrix}$, Inconsistent: 2I + II contradicts III.

(f) $\begin{bmatrix} 1 & 0 & 0 & 10/7 & 11/7 \\ 0 & 1 & 0 & 1/7 & 6/7 \\ 0 & 0 & 1 & -23/14 & 1/7 \end{bmatrix}$, $\frac{1}{7}[11, 6, 1, 0]^t$, $\frac{1}{14}[-20, -2, 23, 2]$.

(j) $\begin{bmatrix} 1 & 0 & 3/4 & 1 & 5/4 \\ 0 & 1 & 1/4 & 0 & -1/4 \\ 0 & 0 & 0 & 0 & 0 \\ 0 & 0 & 0 & 0 & 0 \end{bmatrix}$, $\frac{1}{4}[5, -1, 0, 0]'$, $\frac{1}{4}[-3, -1, 4, 0]'$,

$[-1, 0, 0, 1]'$, III $= 4$I $-$ II, IV $=$ I $+ 2$II. Since this is a rank 2 system with 4 variables, there are two free variables.

(k) $\begin{bmatrix} 1 & 0 & 3/4 & 1 & 0 \\ 0 & 1 & 1/4 & 0 & 0 \\ 0 & 0 & 0 & 0 & 1 \\ 0 & 0 & 0 & 0 & 0 \end{bmatrix}$. Inconsistent. I $+ 2$II contradicts IV.

1.58   A point (x, y) solves the system if and only if it lies on both lines. Since the lines are parallel, there is no solution to the system.

1.2.2 Applications to Circuit Theory

Problems begin on page **45**

SELF-STUDY QUESTIONS

1.4   The new drop is $E = iR = 3 \cdot 7 = 21$ volts.

1.5   The new drop is $E = iR = 2 \cdot 5 = 10$ volts.

1.6   The assumed directions are as in Figure 1.2. We obtain the following equations:
Current Law:

$$i_1 = i_3 + i_2 \quad \text{(Nodes C and F)}$$

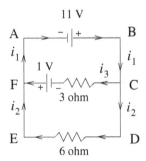

FIGURE 1.2   Assumed flows.

Voltage Law:

$$0 = 11 + 3i_3 + 1 \quad \text{(Loop ABCFA)}$$
$$0 = 3i_3 + 1 - 6i_2 \quad \text{(Loop CFEDC)}$$

EXERCISES

1.60 **(a)** We solve the system found in Exercise 1.6 on page 45. We find $i_1 = \frac{35}{6}$ amp from C to B, $i_2 = \frac{11}{6}$ amp from E to D, $i_3 = 4$ amp from F to C.

(c) The equations are

$$i_1 = i_2 + i_3 \quad \text{(Node C)}$$
$$i_3 = i_4 + i_5 \quad \text{(Node C)}$$
$$0 = 5 + 6i_2 \quad \text{(Loop ABCHA)}$$
$$0 = 6i_2 - 10i_4 - 4 \quad \text{(Loop CHGDC)}$$
$$0 = 10i_4 - 3 - 3i_5 \quad \text{(Loop CHGDC)}$$

yielding the solution $i_1 = -86/15$, $i_2 = -5/6$, $i_3 = -49/10$, $i_4 = -9/10$, $i_5 = -4$.

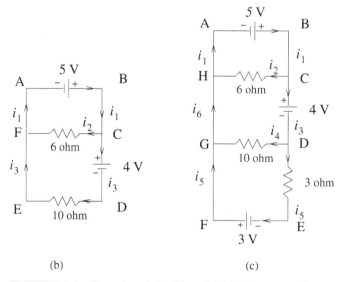

(b)   (c)

FIGURE 1.3   Exercises 1.60.(b) and 1.60.(c) assumed flows.

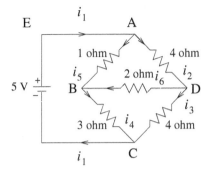

FIGURE 1.4   Exercise 1.61 assumed flows.

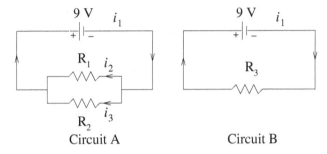

FIGURE 1.5   Exercise 1.62 assumed flows.

1.62   Assuming a clockwise flow of current, the equations are

$$i_1 = i_2 + i_3$$
$$0 = -9 + R_1 i_2$$
$$0 = -9 + R_2 i_3$$

$i_1 = 9\frac{R_2 + R_1}{R_1 R_2}$, $i_2 = 9/R_1$, $i_3 = 9/R_2$.

On the other hand, in Circuit B, the voltage law yields $-9 + i_1 R_3 = 0$ so $i_3 = 9/R_3$, showing the equivalence.

1.3   GAUSSIAN ELIMINATION

Problems begin on page **63**

EXERCISES

1.63   (a) Neither, (b) echelon, (c) neither, (d) echelon (e) reduced echelon.

1.64   We give the solutions followed by the reduced forms.

Solutions:

(a)
$$\begin{bmatrix} 0 \\ -4-t \\ -5/2+t \\ 3/2-1/2\,t \\ t \end{bmatrix},$$

(b)
$$\begin{bmatrix} -1 \\ 1 \\ 0 \\ 1 \end{bmatrix},$$

(c)
$$\begin{bmatrix} -3/2 \\ 5/2 \\ -1/2 \\ 1/2 \end{bmatrix},$$

(d)
$$\begin{bmatrix} 9-2\,t \\ 1 \\ t \\ -2 \\ 3 \end{bmatrix},$$

(e)
$$\begin{bmatrix} 1-2\,t \\ 2 \\ t \\ 1 \\ 3 \end{bmatrix}$$

Reduced forms:

(a)
$$\begin{bmatrix} 1 & 0 & 2 & 4 & 0 & 1 \\ 0 & 1 & -2 & 0 & 3 & 1 \\ 0 & 0 & 2 & 4 & 0 & 1 \\ 0 & 0 & 0 & 2 & 1 & 3 \end{bmatrix},$$

(b)
$$\begin{bmatrix} 0 & 1 & 0 & -2 & 0 \\ 0 & 0 & 1 & 2 & 0 \\ 0 & 0 & 0 & 0 & 1 \end{bmatrix}$$

(c)
$$\begin{bmatrix} 3 & 1 & 2 & 6 & 0 \\ 0 & 2/3 & 1/3 & -1 & 1 \\ 0 & 0 & 2 & 4 & 1 \\ 0 & 0 & 0 & -5/2 & -5/4 \end{bmatrix},$$

(d)
$$\begin{bmatrix} 1 & 1/2 & 0 & 5 & 0 & -1/2 \\ 0 & 0 & 1 & -2 & 0 & 0 \\ 0 & 0 & 0 & 0 & 1 & 1 \end{bmatrix},$$

(e)
$$\begin{bmatrix} 1 & 0 & 2 & 0 & 0 & 1 \\ 0 & 1 & 0 & 0 & 0 & 2 \\ 0 & 0 & 0 & 1 & 0 & 1 \\ 0 & 0 & 0 & 0 & 1 & 3 \end{bmatrix}$$

1.65

(a)
$$\begin{bmatrix} 1 & 0 & 1 & 0 & 3 \\ 0 & 1 & -1 & 0 & 1 \\ 0 & 0 & 0 & 1 & 0 \end{bmatrix},$$

(c)
$$\begin{bmatrix} 1 & 0 & \frac{10}{3} \\ 0 & 1 & \frac{1}{3} \end{bmatrix},$$

(e)
$$\begin{bmatrix} 1 & 0 \\ 0 & 1 \end{bmatrix},$$

(g)
$$\begin{bmatrix} 1 & 0 & -\frac{1}{2} & 0 & 5 \\ 0 & 1 & 1 & 0 & -1 \\ 0 & 0 & 0 & 1 & 2 \end{bmatrix},$$

(j)
$$\begin{bmatrix} 1 & 0 & -1/3 & -1/3 \\ 0 & 1 & 7/3 & 4/3 \\ 0 & 0 & 0 & 0 \\ 0 & 0 & 0 & 0 \end{bmatrix}$$

1.66 Solutions:

(a)
$$\begin{bmatrix} 3-t \\ 1+t \\ t \\ 0 \end{bmatrix},$$

(c)
$$\begin{bmatrix} 10/3 \\ 1/3 \end{bmatrix},$$

(e) represents an inconsistent system.

(g)
$$\begin{bmatrix} 5 + 1/2\,t \\ -1 - t \\ t \\ 2 \end{bmatrix}$$

1.67 We give an echelon form for the coefficient matrix followed by the condition:

(a)
$$\begin{bmatrix} 1 & 1 & 2 & b \\ 0 & -1 & -7 & a - 3\,b \\ 0 & 0 & 0 & c - a - 2b \end{bmatrix}$$

$c = a + 2b,$

(c)
$$\begin{bmatrix} 2 & -3 & -2 & b \\ 0 & 4 & 7 & -2\,b + a \\ 0 & 0 & 0 & c - a \end{bmatrix}$$

$a = c.$

1.68 The right side of I + 2II contradicts III unless $a + 2b = c$.

1.72 We give the coefficient matrix, the reduced form, the solution, and the free variable:

(a)
$$\begin{bmatrix} 2 & 2 & 2 & 3 & 4 \\ 1 & 1 & 1 & 1 & 1 \\ 2 & 3 & 4 & 5 & 2 \\ 1 & 3 & 5 & 11 & 9 \end{bmatrix}, \begin{bmatrix} 1 & 0 & -1 & 0 & 5 \\ 0 & 1 & 2 & 0 & -6 \\ 0 & 0 & 0 & 1 & 2 \\ 0 & 0 & 0 & 0 & 0 \end{bmatrix}$$

$[x, y, z, w]^t = [5 + t, -6 - 2t, t, 2]^t$, free variable: z.

(b)
$$\begin{bmatrix} 2 & 2 & 2 & 3 & 4 \\ 1 & 1 & 1 & 1 & 1 \\ 2 & 4 & 3 & 5 & 2 \\ 1 & 5 & 3 & 11 & 9 \end{bmatrix}, \begin{bmatrix} 1 & 0 & 1/2 & 0 & 2 \\ 0 & 1 & 1/2 & 0 & -3 \\ 0 & 0 & 0 & 1 & 2 \\ 0 & 0 & 0 & 0 & 0 \end{bmatrix},$$

$[x, y, z, w]^t = [2 - t/2, t, -3 - t/2, 2]^t$, free variable: y.

(c) In (a), $t = z$ while in (b) $z = -3 - t/2$. Hence in (a) we replace t with $-3 - t/2$ and simplify to get (b). In (b), $t = y$ while in (a) $y = -6 - 2t$. Hence in (b) we replace t with $-6 - 2t$ and simplify to get (a).

1.74 (a) For a line we want one free variable, hence four pivot variables. Thus we need a rank four system of five equations in five unknowns with no zero coefficients. To create such a system, begin with a 5×6 matrix that is in echelon form which has exactly four non-zero rows and as few non-zero entries as possible and perform elementary row operations on it until a matrix with all non-zero entries is obtained.

(b) For a plane we need two free variables, hence three pivot variables. Thus we need a rank three system of five equations in five unknowns with no zero coefficients. We produce the system exactly as described in (a), except we begin with matrix that is in echelon form which has exactly three non-zero rows.

(c) For a line we want one free variable, hence two pivot variables and rank 2. We produce the example as in part (a) beginning with a 5×3 rank 2 echelon form matrix.

(d) Now we make each equation a multiple of one single equation.

1.75 (a) Since $T = 0$, each spanning vector is the solution obtained by setting one free variable equal to 1 and the other equal to 0. Hence the free variables are z and w.

(b) $\begin{bmatrix} 1 & 0 & 3 & -1 & 0 \\ 0 & 1 & -4 & -4 & 0 \\ 0 & 0 & 0 & 0 & 0 \end{bmatrix}$

1.78 In each part we give the augmented matrix corresponding to the system $B = x_1 X_1 + x_2 X_2 + x_3 X_3$, its reduced form, and our conclusion, which is based upon whether or not the system is consistent.

(a) $\begin{bmatrix} 1 & 1 & 1 & 3 \\ 0 & 2 & 1 & 2 \\ -1 & 1 & 1 & 1 \end{bmatrix}, \begin{bmatrix} 1 & 0 & 0 & 1 \\ 0 & 1 & 0 & 0 \\ 0 & 0 & 1 & 2 \end{bmatrix}$, in the span,

(b) $\begin{bmatrix} 1 & 1 & 1 & a \\ 0 & 2 & 1 & b \\ -1 & 1 & 1 & c \end{bmatrix}, \begin{bmatrix} 1 & 0 & 0 & -c/2+a/2 \\ 0 & 1 & 0 & -c/2-a/2+b \\ 0 & 0 & 1 & c+a-b \end{bmatrix}$, in the span,

(c) $\begin{bmatrix} 1 & 1 & 1 & 3 \\ 0 & 2 & 1 & 2 \\ -1 & 1 & 1 & 1 \end{bmatrix}, \begin{bmatrix} 1 & 0 & 0 & 1 \\ 0 & 1 & 0 & 0 \\ 0 & 0 & 1 & 2 \end{bmatrix}$, not in the span.

1.80 This is not hard. Just pick a vector at random. The chances are that it won't be in the span. To prove it, reason as in Exercise 1.78.

1.82 Since there are n variables and at most $n - 1$ non-zero rows in the reduced form, there must be a free variable.

1.83
$$\begin{bmatrix} 1 & 0 \\ 0 & 1 \end{bmatrix}, \quad \begin{bmatrix} 1 & a \\ 0 & 0 \end{bmatrix}, \quad \begin{bmatrix} 0 & 1 \\ 0 & 0 \end{bmatrix}$$

1.85 (a) From the answer to 1.5, No. (b) It is the 3×3 identity augmented by a column of constants, (c) No. The reduced form has at most two non-zero rows; hence two pivot entries. If consistent, it has a free variable; hence an infinite number of solutions.

1.87 Our system is equivalent with

$$z + w = -x = -s$$
$$2z + w = -y = -t$$

which yields $[x, y, z, w]^t = s[1, 0, 1, -2]^t + t[0, 1, -1, 1]^t$.

1.3.2 Applications to Traffic Flow

Problems begin on page **73**

SELF-STUDY QUESTIONS

1.7 $t + s = 350$.

1.8 $z + v = 450$.

1.9 $90 + x = y$.

EXERCISES

1.90 (a)   The equations are $x + w = 20$, $z + w = 50$, $x + y = 50$, $z + y = 80$, and $x + y + z + w = 100$. We list the variables in the order $[x, y, z, w]^t$. The corresponding matrix and its reduced form are respectively:

$$\begin{bmatrix} 1 & 0 & 0 & 1 & 20 \\ 0 & 0 & 1 & 1 & 50 \\ 1 & 1 & 0 & 0 & 50 \\ 0 & 1 & 1 & 0 & 80 \\ 1 & 1 & 1 & 1 & 100 \end{bmatrix}, \begin{bmatrix} 1 & 0 & 0 & 1 & 20 \\ 0 & 1 & 0 & -1 & 30 \\ 0 & 0 & 1 & 1 & 50 \\ 0 & 0 & 0 & 0 & 0 \\ 0 & 0 & 0 & 0 & 0 \end{bmatrix}$$

The general solution is $[x, y, z, w]^t = [20 - w, 30 + w, 50 - w, w]$

1.4 COLUMN SPACE AND NULLSPACE

Problems begin on page **86**

EXERCISES

1.93 **(a)** $[0, 5, -11]'$,

  (b) $[7, 10, 7, 5]'$,

  (c) $[x_1 + 2x_2 + 3x_3, 4x_1 + 5x_2 + 6x_3]'$.

1.95 Compute AX for the general element X in the appropriate $\mathbb{R}^n$ and set each entry of the result equal to a constant. For (c), for example, you might choose

$$x + 2y + 3z = 17$$
$$4x + 5y + 6z = -4$$

1.96 The nullspace of a matrix A is found by computed by augmenting A with a column of zeros and computing its reduced form R which is the reduced form of A augmented with a column of zeros. Each basis element of the nullspace is found by setting one of the free vectors in the system corresponding to R equal to one and the other free variables equal to 0. In each part we give R and the basis for the nullspace.

(a) $\begin{bmatrix} 1 & 0 & 0 & -4/7 & 0 \\ 0 & 1 & 0 & -3/2 & 0 \\ 0 & 0 & 1 & -6/7 & 0 \end{bmatrix}$, $\{[4/7, 3/2, 6/7, 1]'\}$,

(b) $\begin{bmatrix} 1 & 0 & 0 \\ 0 & 1 & 0 \\ 0 & 0 & 0 \\ 0 & 0 & 0 \end{bmatrix}$, $\{[0, 0]'\}$,

(c) $\begin{bmatrix} 1 & 0 & -1 \\ 0 & 1 & 2 \end{bmatrix}$, $\{[1, -2, 1]'\}$

1.97 The nullspace is spanned by:

  (a) $\{[-1, 1, 1, 0, 0]', [-3, -1, 0, 0, 1]'\}$,

  (c) $\{[-10, -1, 3]'\}$,

(e) $\{[0, 0]\}$,

(g) $\{[\frac{1}{2}, -1, 1, 0, 0]', [-5, 1, 0, -2, 1]'\}$.

1.99 According to Theorem 1.8 the columns of B must be scalar multiples of $[1, 2, 3]'$.

1.100 Let the columns of B be any four vectors which span the same space as the given vectors. For example letting $X = [1, 0, 1]'$ and $Y = [1, 1, 1]'$, we could let $B = [X + Y, X + 2Y, X + 3Y, X + 4Y]$. These vectors span the same space as X and Y since $X = (X + 2Y) - (X + Y)$ and $Y = (X + 2Y) - (X + Y)$.

1.103 The zero vector is always a solution. There are an infinity of solutions due to the more unknowns theorem.

1.104 The reduced form R of A is the matrix in (1.20) on page 37 with its fifth column deleted:

$$R = \begin{bmatrix} 1 & 1 & 1 & 1 \\ 0 & 1 & -1 & -1 \\ 0 & 0 & 0 & 0 \\ 0 & 0 & 0 & 0 \end{bmatrix}$$

The nullspace of A is found by augmenting A with a column of zeros and computing its reduced form R' which is R augmented with a column of zeros.

$$R' = \begin{bmatrix} 1 & 1 & 1 & 1 & 0 \\ 0 & 1 & -1 & -1 & 0 \\ 0 & 0 & 0 & 0 & 0 \\ 0 & 0 & 0 & 0 & 0 \end{bmatrix}$$

The fee variables are z and w. Each basis element of the nullspace is found by setting one of the free vectors in the system corresponding to R equal to one and the other free variables equal to 0. We find that the nullspace is the span of $[-2, 1, 1, 0]'$ and $[-2, 1, 0, 1]'$. This exercise demonstrates the translation theorem.

1.105 (c) One checks by substitution that $[-1, 1, 2, 1, 1, 1]'$ is a particular solution to the non-homogeneous system. Thus the translation theorem says that the expression in (c) is the general solution to the non-homogenous system.

1.106 **(a)** Let the equation be $ax + by + cz = d$. Since 0 belongs to the span, the zero vector solves the equation, showing that $d = 0$. Substituting $[1, 2, 1]^t$ and $[1, 0, -3]^t$ into the equation yields the system

$$a + 2b + \quad c = 0$$
$$a \quad\quad - 3c = 0$$

One solution is $c = 1$, $a = 3$, $b = -2$.

(b) Let each equation be a multiple of the one from (a).

(c) Substitute $[1, 1, 1]^t$ into the system found in (b) producing a vector B. The desired system is $AX = B$.

1.108 True. $Y_1 = 2X_1 + 2X_2$, $Y_2 = X_1 - X_2$. Hence, Y_1 and Y_2 belong to the span of the X_i. Since spans are subspaces, the span of the Y_i is contained in the span of the X_i. Conversely, $X_1 = \frac{1}{4}Y_1 + \frac{1}{2}Y_2$ and $X_2 = \frac{1}{4}Y_1 - \frac{1}{2}Y_2$ showing that the span of the X_i is contained in the span of the Y_i. Hence, the spans are equal.

1.110 No, the two answers are not consistent. If the answers were consistent, then the difference of any two solutions to the system would be a solution to the homogeneous system which, form Group I's, answer is spanned by $[-3, 1, 1]^t$ and $[-1, 0, 1]^t$. Thus, there should exist s and t such that the following equation is true.

$$[1, 0, 0]^t - [1, -1, 1]^t = s[-3, 1, 1]^t + t[-1, 0, 1]^t$$

This system is equivalent with

$$3s + t = 0$$
$$1 - s = 0$$
$$-1 - s - t = 0$$

The last two equations imply $s = 1$ and $t = 2$ which contradicts the first equation, showing the inconsistency.

1.112 **(a)** If W belongs to span $\{X, Y, Z\}$, then

$$W = aX + bY + cZ$$
$$= aX + bY + c(2X + 3Y)$$
$$= (a + 2c)X + (b + 3c)Y$$

which belongs to span $\{X, Y\}$.

Conversely, if W belongs to span $\{X, Y\}$, then

$$W = aX + bY = aX + bY + 0Z$$

which belongs to span $\{X, Y, Z\}$. Thus, the two sets have the same elements and are therefore equal.

(b) From (a) it suffices to prove span $\{X, Y\} =$ span $\{X, Z\}$. This follows from the observations that $Z = 2X + 3Y$ and $Y = \frac{1}{3}Z - \frac{2}{3}X$. Hence, if $Z \in$ span $\{X, Y\}$,

$$
\begin{aligned}
Z &= aX + bY \\
&= aX + b\left(\frac{1}{3}Z - \frac{2}{3}X\right) \\
&= \left(a - \frac{2b}{3}\right)X + \frac{b}{3}Z
\end{aligned}
$$

which belongs to span $\{X, Z\}$.

Conversely, if $Z \in$ span $\{X, Z\}$,

$$
\begin{aligned}
Z &= aX + bZ \\
&= aX + b(2X + 3Y) \\
&= (a + 2b)X + (a + 3b))Z
\end{aligned}
$$

which belongs to span $\{X, Y\}$. Thus, the two sets have the same elements and are therefore equal.

1.115 (c) We prove that $\mathcal{W}$ satisfies the subspace properties. (Theorem 1.13 on page 83, properties 1–3.)

1. Let X and Y be elements of $\mathcal{W}$. Then

$$
X = \begin{bmatrix} a & b \\ c & d \end{bmatrix} \qquad Y = \begin{bmatrix} a' & b' \\ c' & d' \end{bmatrix}
$$

where

$$a + b + c + d = 0 = a' + b' + c' + d'$$

Then

$$
X + Y = \begin{bmatrix} a + a' & b + b' \\ c + c' & d + d' \end{bmatrix}
$$

which belongs to $\mathcal{W}$ since

$$a + a' + b + b' + c + c' + d + d'$$
$$= (a + b + c + d) + (a' + b' + c' + d') = 0$$

2. If X is as above and k is a scalar, then,

$$0 = k(a + b + c + d) = ka + kb + kc + kd$$

which is equivalent with kX belonging to $\mathcal{W}$.

3. Clearly, $\mathcal{W}$ contains the zero vector.

1.116 $\mathcal{W}$ is the first quadrant in $\mathbb{R}^2$. No: $\mathcal{W}$ is not closed under scalar multiplication.

1.118 If the first entry of either X or Y is zero, then $X + Y$ will belong to $\mathcal{W}$. Otherwise, it will not belong to $\mathcal{W}$.

1.119 In this exercise it is easier to use Definition 1.16 on page 77. Start by noting that the general upper triangular matrix is

$$A = \begin{bmatrix} a & b & c \\ 0 & d & e \\ 0 & 0 & f \end{bmatrix}$$

Letting all of the variables equal zero proves that 0 is upper-triangular; hence $\mathcal{T}$ is non-empty.

Let A' be another element of $\mathcal{T}$,

$$A' = \begin{bmatrix} a' & b' & c' \\ 0 & d' & e' \\ 0 & 0 & f' \end{bmatrix}$$

Then for scalars s and t

$$sA + tB = \begin{bmatrix} sa & sb & sc \\ 0 & sd & se \\ 0 & 0 & sf \end{bmatrix} + \begin{bmatrix} ta' & tb' & tc' \\ 0 & td' & te' \\ 0 & 0 & tf' \end{bmatrix}$$

$$= \begin{bmatrix} sa + ta' & sb + tb' & sc + tc' \\ 0 & sd + td' & se + te' \\ 0 & 0 & sf + tf' \end{bmatrix}$$

Hence $sA + tA'$ is upper-triangular, showing that $\mathcal{T}$ is closed under linear combinations; hence a subspace.

1.121 $X = aX + bY$ is a solution if and only if $a + b = 1$.

1.122 **(b)** Suppose that y and z are two solutions. Then

$$y'' + 3y' + 2y = 0$$
$$z'' + 3z' + 2z = 0$$

If we add these two equations, we get

$$(y + z)'' + (y + z)' + 2(y + z) = 0$$

showing that $y + z$ is a solution. Multiplying by c yields

$$(cy)'' + (cy)' + 2cy = 0$$

showing that cy is a solution. The other parts are similar.

1.123 **(b)** We show that none of the subspace properties (Theorem 1.13 on page 83, properties 1–3) holds.

    1. Suppose that

$$y'' + 3y' + 2y = 2t$$
$$z'' + 3z' + 2z = 2t$$

Adding these equations shows

$$\begin{aligned}
(y + z)'' + 3(y + z)' + 2(y + z) &= (y'' + 3y' + 2y) \\
&\quad + (z'' + 3z' + 2z) \\
&= 2t + 2t = 4t \\
&\neq 2t
\end{aligned}$$

Hence property 1 fails.

    2. Also if y is as above and $c \neq 1$. Then

$$(cy)'' + 3(cy)' + 2(cy) = c(y'' + 3y' + 2y) = 2ct \neq 2t$$

Hence property 1 fails.

3.

$$0'' + 30' + 20 = 0 \neq 2t$$

Hence 0 is not in $\mathcal{W}$ and property 3 fails.

1.126 If f and g satisfy $f(1) = g(1) = 0$ then for scalars s and t, $sf(1) + tg(1) = 0$ showing that $\mathcal{W}$ is closed under linear combinations.

1.130 (a) $f(x) = x - 3/2$.

(b) This is clear since for scalars s and t,

$$\int_1^2 (sf(x) + tg(x)) \, dx = s \int_1^2 f(x) \, dx + t \int_1^2 g(x) \, dx$$

(c) From the preceding formula for $f, g \in \mathcal{V}$, $sf + tg \in \mathcal{V}$ if and only if $s + t = 1$. Hence $\mathcal{V}$ is not a subspace.

1.132 $S \cup T$ is a subspace only if either $S \subset T$ or $T \subset S$. For the proof, suppose that S is not contained in T. Then S contains an element S which is not in T. Then for all T in T, $U = S + T$ must be in $S \cup T$. But U cannot be in T since $S = T - U$ and S is not in T. Thus, U belongs to S, proving that $T = U - S$ belongs to S. Hence, $T \subset S$.

CHAPTER 2

LINEAR INDEPENDENCE AND DIMENSION

2.1 THE TEST FOR LINEAR INDEPENDENCE

Problems begin on page **108**

EXERCISES

2.1 Let the matrices in each part be A_i, $i = 1, \ldots$. We provide the relations among the A_i, the coefficient matrix for the dependency equation, and its reduced form R. The most efficient way of finding the relations between the A_i is to use Theorem 2.5 on page 106 which implies that the relations between the A_i are the same as the relations between the columns of R.

(a) independent, $M = [A_1, A_2, A_3]$, $R = I$.

(d) independent, $M = \begin{bmatrix} 2 & 1 & 17 & 0 \\ 3 & 3 & 0 & 5 \\ 0 & 0 & 9 & 0 \\ 1 & 0 & 1 & 6 \end{bmatrix}$, $R = I$.

Solutions Manual to Accompany Linear Algebra: Ideas and Applications, Fourth Edition. Richard Penney.
© 2016 John Wiley & Sons, Inc. Published 2016 by John Wiley & Sons, Inc.

(g) dependent, $A_3 = -3A_1 + 4A_2$, $M = \begin{bmatrix} 1 & 2 & 5 \\ 2 & 1 & -2 \\ 3 & 3 & 3 \\ 2 & 4 & 10 \end{bmatrix}$ $R = \begin{bmatrix} 1 & 0 & -3 \\ 0 & 1 & 4 \\ 0 & 0 & 0 \\ 0 & 0 & 0 \end{bmatrix}$

(j) independent, $M = [A_1^t, A_2^t, A_3^t]$ $R = I$

2.2 (b) $M(2,2)$.

2.3 Let the columns of A be $A_1, \ldots, A_n$ The coefficient matrix for dependency equation $x_1 A_1 + \cdots + x_n A_n = 0$ is A. We compute the row reduced form of A and find a basis for the solution set of the system $AX = 0$. Replacing the x_i in the dependency equation with the corresponding entry in any one of the basis elements yields one relation between the A_j. For example in (a) the first basis element is $[-3, 3, 1, 0]^t$ and the corresponding relation is $-3A_1 + 3A_2 + A_3 = 0$. For each problem, we give the reduced echelon form of A, a basis for the solution set, and the corresponding relations.

(a) $-3A_1 + 3A_2 + A_3 = 0$, $-\frac{5}{3}A_1 - \frac{2}{3}A_2 + A_4 = 0$,

$$\begin{bmatrix} 1 & 0 & 3 & 5/3 \\ 0 & 1 & -3 & 2/3 \\ 0 & 0 & 0 & 0 \\ 0 & 0 & 0 & 0 \end{bmatrix}, \begin{bmatrix} -3 \\ 3 \\ 1 \\ 0 \end{bmatrix}, \begin{bmatrix} -5/3 \\ -2/3 \\ 0 \\ 1 \end{bmatrix}$$

(c) $\frac{1}{3}A_1 + \frac{8}{3}A_2 + A_3 = 0$

$$\begin{bmatrix} 1 & 0 & -1/3 \\ 0 & 1 & -8/3 \\ 0 & 0 & 0 \\ 0 & 0 & 0 \\ 0 & 0 & 0 \end{bmatrix}, \begin{bmatrix} 1/3 \\ 8/3 \\ 1 \end{bmatrix}$$

(e) $-\frac{2}{5}A_1 - \frac{23}{10}A_2 - \frac{19}{5}A_3 + A_4 = 0$

$$\begin{bmatrix} 1 & 0 & 0 & \frac{2}{5} \\ 0 & 1 & 0 & \frac{23}{10} \\ 0 & 0 & 1 & \frac{19}{5} \\ 0 & 0 & 0 & 0 \end{bmatrix}, \begin{bmatrix} -\frac{2}{5} \\ -\frac{23}{10} \\ -\frac{19}{5} \\ 1 \end{bmatrix}$$

2.5 Let A_i be the columns of A. From the reduced form R of A and Theorem 2.5 on page 106, A_1 and A_2 are a basis of the column space. The relations between the A_i are the same as those between the columns R_i of R. Since $R_3 = R_1 + R_2$ and $R_4 = R_1 - 2R_2$, we see $A_3 = A_1 + A_2$ and $A_4 = A_1 - 2A_2$.

$$R = \begin{bmatrix} 1 & 0 & 1 & 1 \\ 0 & 1 & 1 & -2 \\ 0 & 0 & 0 & 0 \\ 0 & 0 & 0 & 0 \end{bmatrix}$$

2.6 Each vector has a nonzero entry in the positions, where the other two vectors have zeros.

2.7 Let the rows of A be A_1, A_2, and A_3. Then $xA_1 + yA_2 + zA_3 = 0$ yields the system

$$\begin{aligned} x &= 0 & xc + yf &= 0 \\ xa &= 0 & xd + yg + z &= 0 \\ xb + y &= 0 & xe + yh + zk &= 0 \end{aligned}$$

The first, third, and fifth equations show that $x = y = z = 0$, proving independence.

2.9 **(a)** The result is that if A be a 4×3 matrix with linearly independent columns its reduced form is the matrix R below.

$$R = \begin{bmatrix} 1 & 0 & 0 \\ 0 & 1 & 0 \\ 0 & 0 & 1 \\ 0 & 0 & 0 \end{bmatrix}$$

For the proof note that since none of the columns is a combination of other columns, every column is a pivot column. Hence the reduced form must be as stated.

(b) The result is that if A is a 4×7 matrix whose first three columns are linearly independent, then the first three columns must be pivot columns. However there could be other pivot columns. For example, two possible reduced forms of A might be

$$\begin{bmatrix} 1 & 0 & 0 & a & b & c & d \\ 0 & 1 & 0 & e & f & g & h \\ 0 & 0 & 1 & i & j & k & l \\ 0 & 0 & 0 & 0 & 0 & 0 & 0 \end{bmatrix}$$

or

$$\begin{bmatrix} 1 & 0 & 0 & a & 0 & b & c \\ 0 & 1 & 0 & d & 0 & e & f \\ 0 & 0 & 1 & g & 0 & h & i \\ 0 & 0 & 0 & 0 & 1 & j & k \end{bmatrix}$$

There are many other possibilities. For the proof note that since we do not interchange columns during the reduction process, the process of reducing A also reduces $[A_1, A_2, A_3]$. Hence from (a), the first three columns of the reduced form are the columns of R.

2.10 Since there is exactly one pivot entry in each non-zero row of the reduced form, the number of non-zero rows equals the number of pivot columns. Hence we expect the number of basis elements to equal the rank.

2.11 Suppose that

$$x_1 Y_1 + x_2 Y_2 = 0$$

Then substituting the expressions for Y_i yields

$$x_1(3X_1 - 2X_2) + x_2(X_1 + X_2) = 0$$
$$(3x_1 + x_2)X_1 + (-2x_1 + x_2)X_2 = 0$$

It follows from the independence of the X_i that

$$3x_1 + x_2 = 0$$
$$-2x_1 + x_2 = 0$$

The augmented matrix for this system and its reduced form are respectively

$$\begin{bmatrix} 3 & 1 & 0 \\ -2 & 1 & 0 \end{bmatrix} \quad \begin{bmatrix} 1 & 0 & 0 \\ 0 & 1 & 0 \end{bmatrix}$$

Hence $x_1 = x_2 = 0$, proving independence.

2.12 Suppose that

$$x_1 Y_1 + x_2 Y_2 = 0$$

Then substituting the expressions for Y_i yields

$$x_1(aX_1 + bX_2) + x_2(cX_1 + dX_2) = 0$$
$$(ax_1 + cx_2)X_1 + (bx_1 + dx_2)X_2 = 0$$

It follows from the independence of the X_i that

$$ax_1 + cx_2 = 0$$
$$bx_1 + dx_2 = 0$$

This system is equivalent with the vector equation

$$x_1 \begin{bmatrix} a \\ b \end{bmatrix} + x_2 \begin{bmatrix} c \\ d \end{bmatrix} = \begin{bmatrix} 0 \\ 0 \end{bmatrix}$$

which is the dependency equation for the pair of vectors $[a, b]^t$ and $[c, d]^t$, proving the result.

2.15 We need to show that $\{[1, 1, 0]^t, [1, 2, 0]^t, [1, 1, 1]^t\}$ is linearly independent in $\mathbb{R}^3$. The dependency equation is

$$x \begin{bmatrix} 1 \\ 1 \\ 0 \end{bmatrix} + y \begin{bmatrix} 1 \\ 2 \\ 0 \end{bmatrix} + z \begin{bmatrix} 1 \\ 1 \\ 1 \end{bmatrix} = \begin{bmatrix} 0 \\ 0 \\ 0 \end{bmatrix}$$

which is equivalent with the matrix equality

$AX = 0$ where

$$A = \begin{bmatrix} 1 & 1 & 1 \\ 1 & 2 & 1 \\ 0 & 0 & 1 \end{bmatrix}, X = \begin{bmatrix} x \\ y \\ z \end{bmatrix}$$

It is easily seen that the reduced form of A is I. Hence $X = 0$. The independence follows.

2.17 Consider the equation $xY_1 + yY_2 + xY_3 = 0$. Substituting the expressions for Y_i into this equation yields

$$x(X_1 + 2X_2 - X_3) + y(2X_1 + 2X_2 - X_3) + z(4X_1 + 2X_2 - X_3) = 0$$
$$(2x + y + 2z)X_1 + (3x - y + 13z)X_2 + (5x + 3y + 3z)X_3 = 0$$

There are an infinite number of solutions to the system

$$x + 2y + 4z = 0$$
$$2x + 2y + 2z = 0$$
$$-x - y - z = 0$$

since the reduced form of its coefficient matrix is

$$\begin{bmatrix} 1 & 0 & -2 \\ 0 & 1 & 3 \\ 0 & 0 & 0 \end{bmatrix}$$

Hence the Y_i are dependent. We do not need the independence of the X_i.

2.20 The result follows from the observation that from formula (1.42) on page 78 the equation $AX = 0$ is equivalent with the dependency equation for the columns of A.

2.21 Let the vectors be $A = [a_1, a_2]'$, $B = [b_1, b_2]'$, and $C = [c_1, c_2]'$. The equation $xA + yB + zC = 0$ is equivalent with the system

$$a_1x + b_1y + c_1z = 0$$
$$a_2x + b_2y + c_2z = 0$$

which has a non-trivial solution due to the More Unknowns Theorem since the system is clearly consistent.

2.23 Let your matrices be

$$
A = \begin{bmatrix} a_1 \\ a_2 \\ a_3 \\ a_4 \end{bmatrix}, \;
B = \begin{bmatrix} b_1 \\ b_2 \\ b_3 \\ b_4 \end{bmatrix}, \;
C = \begin{bmatrix} c_1 \\ c_2 \\ c_3 \\ c_4 \end{bmatrix}, \;
D = \begin{bmatrix} d_1 \\ d_2 \\ d_3 \\ d_4 \end{bmatrix}, \;
E = \begin{bmatrix} e_1 \\ e_2 \\ e_3 \\ e_4 \end{bmatrix}
$$

Then the equation $x_1 A + x_2 B + x_3 C + x_4 D + x_5 E = 0$ results in the system

$$
\begin{aligned}
a_1 x_1 + b_1 x_2 + c_1 x_3 + d_1 x_4 + e_1 x_5 &= 0 \\
a_2 x_1 + b_2 x_2 + c_2 x_3 + d_2 x_4 + e_2 x_5 &= 0 \\
a_3 x_1 + b_3 x_2 + c_3 x_3 + d_3 x_4 + e_3 x_5 &= 0 \\
a_4 x_1 + b_4 x_2 + c_4 x_3 + d_4 x_4 + e_4 x_5 &= 0
\end{aligned}
$$

Since this is a system four homogeneous equations in five unknowns, it has a nonzero solution due to the More Unknowns Theorem, proving dependence.

2.24 **(b)** $\cos(2x) = \cos^2 x - \sin^2 x$

  (d) $\cos(2x) + 2\sin^2 x = 1$

  (e) $\ln(3x) = \ln x + (\ln 3) \cdot 1.$

2.25 For each problem we give the coefficient matrix from formula (2.9) on page 107 and its value at a specific value of x. This matrix is the coefficient matrix for the dependency equation and its derivatives, evaluated at this point. The proof amounts to reducing the latter matrix to get a matrix of maximum rank.

(b)
$$
\begin{bmatrix}
e^x & e^{2x} & e^{3x} \\
e^x & 2e^{2x} & 3e^{3x} \\
e^x & 4e^{2x} & 9e^{3x}
\end{bmatrix}
x = 0 :
\begin{bmatrix}
1 & 1 & 1 \\
1 & 2 & 3 \\
1 & 4 & 9
\end{bmatrix}
$$

(e)
$$
\begin{bmatrix}
e^x & \sin(2x) & e^x \sin(2x) \\
e^x & 2\cos(2x) & e^x \sin(2x) + 2e^x \cos(2x) \\
e^x & -4\sin(2x) & -3e^x \sin(2x) + 4e^x \cos(2x)
\end{bmatrix}
x = 0 :
\begin{bmatrix}
1 & 0 & 0 \\
1 & 2 & 2 \\
1 & 0 & 4
\end{bmatrix}
$$

(g)
$$
\begin{bmatrix}
\ln x & x \ln x \\
x^{-1} & \ln x + 1 \\
-x^{-2} & x^{-1}
\end{bmatrix}
x = 1 :
\begin{bmatrix}
0 & 0 \\
1 & 1 \\
-1 & 1
\end{bmatrix}
$$

2.2 DIMENSION

Problems begin on page **123**

EXERCISES

2.28 Let $B = [a, b]^t$ where a and b are specific numbers. The equation $B = xA_1 + yA_2$ is equivalent with the system

$$x + 2y = a$$
$$2x - 3y = b$$

The augmented matrix and its reduced form are

$$\begin{bmatrix} 1 & 2 & a \\ 2 & 3 & b \end{bmatrix} \quad \begin{bmatrix} 1 & 0 & 2b - 3a \\ 0 & 1 & -b + 2a \end{bmatrix}$$

Then $B = (2b - 3a)A_1 + (-b + 2a)A_2$. For example, if $B = [2, 7]^t$ then $x = 8$ and $y = -3$.

2.30 B can be any vector not of the form $[c, 2c]$ and C must be of this form.

2.31 **(a)** Yes: $Z = 2X_1 + 3X_2$.

(b) No. The augmented matrix M for the system defined by $aX_1 + bA_2 = [1, 2, 3]^t$ and its reduced form R is shown below. It follows that the system is inconsistent.

$$M = \begin{bmatrix} 1 & 3 & 1 \\ 2 & -7 & 2 \\ -3 & 2 & 3 \end{bmatrix}, R = \begin{bmatrix} 1 & 0 & 0 \\ 0 & 1 & 0 \\ 0 & 0 & 1 \end{bmatrix}$$

2.33 **(b)** $\mathcal{W}$ is a subspace because it is a span. A basis is

$$\begin{bmatrix} 1 & 1 \\ 2 & 1 \end{bmatrix}, \quad \begin{bmatrix} 1 & 1 \\ 0 & 1 \end{bmatrix}$$

since

$$\begin{bmatrix} a+b+2c+3d & a+b+2c+3d \\ 2a+2c+4d & a+b+2c+3d \end{bmatrix}$$

$$= a\begin{bmatrix} 1 & 1 \\ 2 & 1 \end{bmatrix} + b\begin{bmatrix} 1 & 1 \\ 0 & 1 \end{bmatrix} + c\begin{bmatrix} 2 & 2 \\ 2 & 2 \end{bmatrix} + d\begin{bmatrix} 3 & 3 \\ 4 & 3 \end{bmatrix}$$

and

$$\begin{bmatrix} 2 & 2 \\ 2 & 2 \end{bmatrix} = \begin{bmatrix} 1 & 1 \\ 2 & 1 \end{bmatrix} + \begin{bmatrix} 1 & 1 \\ 0 & 1 \end{bmatrix}$$

$$\begin{bmatrix} 3 & 3 \\ 4 & 3 \end{bmatrix} = 2\begin{bmatrix} 1 & 1 \\ 2 & 1 \end{bmatrix} + \begin{bmatrix} 1 & 1 \\ 0 & 1 \end{bmatrix}$$

2.34 **(b)** We reason as in Example 2.7 on page 118.

2.35 **(b)** The dimension is 6. The general element of the space is

$$\begin{bmatrix} a & b & c \\ b & d & e \\ c & e & f \end{bmatrix} = a\begin{bmatrix} 1 & 0 & 0 \\ 0 & 0 & 0 \\ 0 & 0 & 0 \end{bmatrix} + b\begin{bmatrix} 0 & 1 & 0 \\ 1 & 0 & 0 \\ 0 & 0 & 0 \end{bmatrix} + c\begin{bmatrix} 0 & 0 & 1 \\ 0 & 0 & 0 \\ 1 & 0 & 0 \end{bmatrix}$$

$$+ d\begin{bmatrix} 0 & 0 & 0 \\ 0 & 1 & 0 \\ 0 & 0 & 0 \end{bmatrix} + e\begin{bmatrix} 0 & 0 & 0 \\ 0 & 0 & 1 \\ 0 & 1 & 0 \end{bmatrix} + f\begin{bmatrix} 0 & 0 & 0 \\ 0 & 0 & 0 \\ 0 & 0 & 1 \end{bmatrix}$$

The set consisting of the six matrices on the right is the basis.

(c) The dimension is 1. The general element of the space is

$$\begin{bmatrix} 0 & a \\ -a & 0 \end{bmatrix} = a\begin{bmatrix} 0 & 1 \\ -1 & 0 \end{bmatrix}$$

The set consisting of the matrix on the right is the basis.

(d) The dimension is 6. The general element of the space is

$$\begin{bmatrix} a & b & c \\ 0 & d & e \\ 0 & 0 & f \end{bmatrix} = a\begin{bmatrix} 1 & 0 & 0 \\ 0 & 0 & 0 \\ 0 & 0 & 0 \end{bmatrix} + b\begin{bmatrix} 0 & 1 & 0 \\ 0 & 0 & 0 \\ 0 & 0 & 0 \end{bmatrix} + c\begin{bmatrix} 0 & 0 & 1 \\ 0 & 0 & 0 \\ 0 & 0 & 0 \end{bmatrix}$$

$$+ d\begin{bmatrix} 0 & 0 & 0 \\ 0 & 1 & 0 \\ 0 & 0 & 0 \end{bmatrix} + e\begin{bmatrix} 0 & 0 & 0 \\ 0 & 0 & 1 \\ 0 & 0 & 0 \end{bmatrix} + f\begin{bmatrix} 0 & 0 & 0 \\ 0 & 0 & 0 \\ 0 & 0 & 1 \end{bmatrix}$$

The set consisting of the six matrices on the right is the basis.

2.37 We give the basis where the original vectors are denoted $A_i, i = 1, ...,$ as well as the coefficient matrix for the dependency equation and it's reduced form.

(a) Basis: $\{A_1, A_2, A_4\}$

$$\begin{bmatrix} 5 & -2 & 4 & 5 \\ -3 & 1 & -3 & 1 \\ 2 & -1 & 1 & -4 \\ 4 & -2 & 2 & -8 \end{bmatrix} \quad \begin{bmatrix} 1 & 0 & 2 & 0 \\ 0 & 1 & 3 & 0 \\ 0 & 0 & 0 & 1 \\ 0 & 0 & 0 & 0 \end{bmatrix}$$

(c) Basis: $\{A_1, A_2, A_3\}$

$$\begin{bmatrix} 1 & 2 & 1 \\ 2 & 5 & 2 \\ -2 & -1 & 0 \\ 3 & -1 & -1 \end{bmatrix} \quad \begin{bmatrix} 1 & 0 & 0 \\ 0 & 1 & 0 \\ 0 & 0 & 1 \\ 0 & 0 & 0 \end{bmatrix}$$

2.38 We find a basis for the nullspace by solving the equation $AX = 0$. Specifically we augment A with a null column and reduce. In each case, if A is $n \times m$ then rank $= n -$ nullity. We give the rank, the reduced form of the augmented matrix, and the elements of the basis.

(b) rank: 2, $\begin{bmatrix} 1 & 0 & 1 & 2/5 & 0 \\ 0 & 1 & 0 & 1/5 & 0 \\ 0 & 0 & 0 & 0 & 0 \end{bmatrix}$, $\begin{bmatrix} -2/5 \\ -1/5 \\ 0 \\ 1 \end{bmatrix}$, $\begin{bmatrix} -1 \\ 0 \\ 1 \\ 0 \end{bmatrix}$.

2.41 **(a)** Any vector which is independent of X_1 and X_2 will work. Choose X_3 randomly and test $\{X_1, X_2, X_3\}$ for independence.

(b) There must exist a vector X_3 which is not in the span of X_1 and X_2 since these vectors span a two dimensional subspace. It follows as in the proof of Theorem 2.8 that then $\{X_1, X_2, X_3\}$ is independent and hence forms a basis.

2.43 **(a)** It is easily proven that the X_i are independent; hence span the nullspace.

(b) $[1, 1, 1, 1]^t = \frac{1}{4}X_1 + \frac{1}{4}X_2 + \frac{1}{4}X_3$, showing that $[1, 1, 1, 1]^t$ belongs to the nullspace.

2.45 Since A and B span, C is a linear combination of A and B; hence $\{A, B, C\}$ is dependent. $\{B, C\}$ need not be a basis of $\mathcal{W}$, for example C could be a multiple of B.

2.49 If $\mathcal{W} = \{0\}$ then 0 spans $\mathcal{W}$. Hence assume $\mathcal{W} \neq \{0\}$. From Theorem 2.6 on page 115, $\mathcal{W}$ can contain at most n linearly independent elements. Let $S = \{X_1, X_2, \ldots, X_k\}$ be a linearly independent subset of $\mathcal{W}$, where k is as large as possible and let $W \in \mathcal{W}$. Then by hypothesis $\{W, X_1, \ldots, X_n\}$ is linearly dependent. It follows exactly as in the proof of Theorem 2.8 on page 119 that W is a linear combination of the X_i. (You should supply the details.) Hence the set of X_i spans $\mathcal{W}$.

2.53 **(a)** The entries of X can be any sequence whose limit is zero. For example

$$X = [1/1/2, \ldots, 1/n, \ldots$$

or

$$X = [1/1/4, \ldots, 1/n^2, \ldots$$

etc.

The entries of Y can be any sequence whose limit is not zero. For example

$$Y = [1, 1, \ldots, 1, \ldots$$

or

$$Y = [-1, 1, -1, 1, \ldots, (-1)^n, \ldots$$

or

$$Y = [1/2, 2/3, \ldots, n/(n+1), \ldots$$

(b) Let X and Y be elements of $\mathcal{W}$ where

$$X = [x_1, x_2, \ldots, x_n, \ldots$$
$$Y = [y_1, y_2, \ldots, y_n, \ldots$$

Then $\lim_{n\to\infty} x_n = 0$ and $\lim_{n\to\infty} y_n = 0$. Hence for all scalars s and t

$$\lim_{n\to\infty}(sx_n + ty_n) = s\lim_{n\to\infty} x_n + t\lim_{n\to\infty} y_n = 0$$

Hence

$$sX + tY = [sx_1, sx_2, \ldots, sx_n, \ldots + [ty_1, ty_2, \ldots, ty_n, \ldots$$
$$= [sx_1 + ty_1, sx_2 + ty_2, \ldots, sx_n + ty_n, \ldots$$

is an element of $\mathcal{W}$ showing that $\mathcal{W}$ is closed under linear combinations. Since from part (a) $\mathcal{W}$ is non-empty, it follows that $\mathcal{W}$ is a subspace.

(c) The proof is the same as the proof that $\mathbb{R}^8$ is infinite dimensional. (You should supply the details.)

(d) $[\{[a, b, c, 0, 0, \ldots, 0, \ldots$.

2.54 The set $\{x^n \mid n \in \mathbb{N}\}$ is an infinite independent set in $\mathcal{P}$. Hence $\mathcal{P}$ cannot be k dimensional for any $k \in \mathbb{N}$.

2.2.2 Applications to Differential Equations

Problems begin on page **131**

SELF-STUDY QUESTIONS

2.1 (a) Any function of the form $Ae^{2t} + Be^{-3t}$ is a solution so make three different choices of (A, B).

(b) Since this is a second order equation from Theorem 2.9 on page 129 the dimension of the solution space is 2.

(c) From Theorem 2.10 on page 129, $\{y_1, y_2\}$ is an independent set and from Theorem 2.9 on page 129 the dimension of the solution space is 2. Hence $\{y_1, y_2\}$ spans the solution space.

(d) Since $p(r) = r^5 - 3r^3 + 2r^2$, the differential equation is $y^{(5)} - 3y''' + 2y'' = 0$.

(e) See the answer to part (a).

2.2 False. From Theorem 2.10 on page 129 the dimension of the solution space is four, so three elements will not span.

EXERCISES

2.59 (a) The characteristic polynomial factors as $(r+1)(r+2)$. The basis is $\{e^{-t}, e^{-2t}\}$. The general solution is $y(t) = ce^{-t} + de^{-2t}$.

(b) The roots of the characteristic polynomial are $-2 \pm 3i$. Hence, there are no solutions of the form e^{rt} where r is real.

(c) The characteristic polynomial factors as $(r+1)^2(r-1)$ so the basis is $\{e^{-t}, te^{-t}, e^t\}$.

2.62 The matrix of derivatives and its values at $t = 0$ are respectively

$$\begin{bmatrix} e^{-2t}\cos(3t) & e^{-2t}\sin(3t) \\ -2e^{-2t}\cos(3t) - 3e^{-2t}\sin(3t) & -2e^{-2t}\sin(3t) + 3e^{-2t}\cos(3t) \end{bmatrix}$$

$$\begin{bmatrix} 1 & 0 \\ -2 & 3 \end{bmatrix}$$

The reduced form of this matrix is I showing the independence.

The general solution is $y(t) = Ae^{-2t}\cos 3t + Be^{-2t}\sin 3t$. The solution $y(t) = e^{-2t}\cos 3t + \frac{4}{3}e^{-2t}\sin 3t$ satisfies $y(0) = 1$ and $y'(0) = 2$.

2.3 ROW SPACE AND THE RANK-NULLITY THEOREM

Problems begin on page 143

EXERCISES

2.64 Let the rows of A be A_i, $i = 1, \ldots, 4$. A row vector $D \in M(1,4)$ belongs to the row space of A if and only if there are scalars x_i, $1 \leq i \leq 4$, such that

$$x_1 A_1 + x_2 A_2 + x_3 A_3 + x_4 A_4 = D$$

Taking transposes we see that this is equivalent with

$$x_1A_1^t + x_2A_2^t + x_3A_3^t + x_4A_4^t = D^t$$

which is equivalent with the equation $A^tX = D^t$ where $X = [x_1, x_2, x_3, x_4]^t$.

This is solvable if and only if the reduced form of the augmented matrix $[A^t|D^t]$ represents a consistent system. For B and C these matrices and their reduced forms are respectively

$$\begin{bmatrix} 2 & 1 & 0 & 3 & 4 \\ 1 & 1 & 1 & 3 & 1 \\ 3 & 3 & 2 & 8 & 2 \\ 1 & 0 & 1 & 2 & 5 \end{bmatrix} \begin{bmatrix} 2 & 1 & 0 & 3 & 1 \\ 1 & 1 & 1 & 3 & 2 \\ 3 & 3 & 2 & 8 & 3 \\ 1 & 0 & 1 & 2 & 4 \end{bmatrix}$$

$$\begin{bmatrix} 1 & 0 & 0 & 1 & 4 \\ 0 & 1 & 0 & 1 & -4 \\ 0 & 0 & 1 & 1 & 1 \\ 0 & 0 & 0 & 0 & 0 \end{bmatrix} \begin{bmatrix} 1 & 0 & 0 & 1 & 0 \\ 0 & 1 & 0 & 1 & 0 \\ 0 & 0 & 1 & 1 & 0 \\ 0 & 0 & 0 & 0 & 1 \end{bmatrix}$$

Hence B belongs to the row space and C does not.

2.66 The matrices and their reduced forms are as follows. The basis for the row space is the set of non-zero rows in the reduced form and the basis for the column space is the set of pivot columns *in the original matrix A*. In each part we give the reduced form of the matrix, the basis for the row space and the basis for the column space.

(a)
$$\begin{bmatrix} 1 & 0 & 0 & -2 \\ 0 & 1 & 0 & 0 \\ 0 & 0 & 1 & 7/2 \end{bmatrix} \begin{bmatrix} 1 & 0 & 0 & -2] \\ [0 & 1 & 0 & 0] \\ [0 & 0 & 1 & 7/2] \end{bmatrix} \begin{bmatrix} 1 \\ 2 \\ 2 \end{bmatrix} \begin{bmatrix} 2 \\ 3 \\ 10 \end{bmatrix} \begin{bmatrix} 0 \\ 2 \\ 4 \end{bmatrix}$$

(c)
$$\begin{bmatrix} 1 & 0 & 1/5 \\ 0 & 1 & 3/5 \\ 0 & 0 & 0 \end{bmatrix} \begin{bmatrix} [1 & 0 & 1/5] \\ [0 & 1 & 3/5] \end{bmatrix} \begin{bmatrix} 2 \\ 1 \\ 5 \end{bmatrix} \begin{bmatrix} 1 \\ 3 \\ 0 \end{bmatrix}$$

2.67 The basis for the row space of a matrix A is the set formed by *taking the transposes* of the pivot columns in A^t. The basis for the column space of A is formed by taking the transposes of the non-zero rows in the reduced form of A^t. In each part we give the reduced form of

the transpose, the basis for the row space of the original matrix, and
the basis for the column space of the original matrix.

(a)
$$\begin{bmatrix} 1 & 0 & 0 \\ 0 & 1 & 0 \\ 0 & 0 & 1 \\ 0 & 0 & 0 \end{bmatrix} \begin{bmatrix} [1 & 2 & 0 & -2] \\ [2 & 3 & 2 & 3] \\ [2 & 10 & 4 & 10] \end{bmatrix} \begin{bmatrix} 1 \\ 0 \\ 0 \end{bmatrix} \begin{bmatrix} 0 \\ 1 \\ 0 \end{bmatrix} \begin{bmatrix} 0 \\ 0 \\ 1 \end{bmatrix}$$

(b)
$$\begin{bmatrix} 1 & 0 & 0 \\ 0 & 1 & 0 \\ 0 & 0 & 1 \end{bmatrix} \begin{bmatrix} [-1 & 4 & -2] \\ [\ 4 & 4 & 2] \\ [\ 3 & 0 & -3] \end{bmatrix} \begin{bmatrix} 1 \\ 0 \\ 0 \end{bmatrix} \begin{bmatrix} 0 \\ 1 \\ 0 \end{bmatrix} \begin{bmatrix} 0 \\ 0 \\ 1 \end{bmatrix}$$

2.70 Let the vectors given in the exercise be denoted V_i, $1 \le i \le 4$.

(a)   We create a matrix A having the given vectors as rows. The basis
is the set of non-zero rows of the reduced form, thought of as
columns. This matrix and its reduced form are respectively

$$\begin{bmatrix} 2 & 3 & 1 & 2 \\ 5 & 2 & 1 & 2 \\ 1 & -4 & -1 & -2 \\ 11 & 0 & 1 & 2 \end{bmatrix} \quad \begin{bmatrix} 1 & 0 & 1/11 & 2/11 \\ 0 & 1 & 3/11 & 6/11 \\ 0 & 0 & 0 & 0 \\ 0 & 0 & 0 & 0 \end{bmatrix}$$

Hence the basis is $\{X_1, X_2\}$ where

$$X_1 = \left[1, 0, \frac{1}{11}, \frac{2}{11}\right]^t$$

$$X_2 = \left[0, 1, \frac{3}{11}, \frac{6}{11}\right]^t$$

The dimension is 2.

(b)   Let $X = [x, y, z, w]^t$ be a vector in $\mathbb{R}^4$ that we hope to be able
to express as a linear combination of X_1 and X_2. The equation
$X = aX_1 + bX_2$ is equivalent with

$$\begin{bmatrix} x \\ y \\ z \\ w \end{bmatrix} = a \begin{bmatrix} 1 \\ 0 \\ 1/11 \\ 2/11 \end{bmatrix} + b \begin{bmatrix} 0 \\ 1 \\ 3/11 \\ 6/11 \end{bmatrix}$$

Hence, if a and b exist, $a = x$ and $b = y$. In particular

$$\begin{aligned}
[2, 3, 1, 2]^t &= 2X_1 + 3X_2 \\
[5, 2, 1, 2]^t &= 5X_1 + 2X_2 \\
[1, -4, -1, -2]^t &= X_1 - 4X_2 \\
[11, 0, 1, 2]^t &= 11X_1
\end{aligned} \qquad (2.1)$$

Remark Expressing a general element X of $\mathbb{R}^4$ and a linear combination of the basis elements was easy because of the property that each basis element has a one in a position where the others have zeros. The coefficients were the entries of X in the position where the corresponding basis element has a one and the others have zeros. Notice, however, that since $\{V_1, V_2\}$ does not span $\mathbb{R}^4$, the equation $X = xV_1 + yV_2$ will hold if and only if X actually belongs to the span. For example, the equation

$$[1, 2, 3, 4]^t = V_1 + 2V_2$$

is not valid showing that $[1, 2, 3, 4]^t$ is not in the span. We did not need to check the validity of the equalities in (2.1) since we knew that the rows were in the row space.

(c) Compute the reduced form R of A^t to determine the pivot columns of A^t. The set of pivot columns of A^t is our basis.

$$A^t = \begin{bmatrix} 2 & 5 & 1 & 11 \\ 3 & 2 & -4 & 0 \\ 1 & 1 & -1 & 1 \\ 2 & 2 & -2 & 2 \end{bmatrix} \qquad R = \begin{bmatrix} 1 & 0 & -2 & -2 \\ 0 & 1 & 1 & 3 \\ 0 & 0 & 0 & 0 \\ 0 & 0 & 0 & 0 \end{bmatrix}$$

Basis: $\{V_1, V_2\}$ where

$$V_1 = \begin{bmatrix} 2 \\ 3 \\ 1 \\ 2 \end{bmatrix}, V_2 = \begin{bmatrix} 5 \\ 2 \\ 1 \\ 2 \end{bmatrix}$$

2.71 We create a matrix A having the transposes of the given vectors as rows:

$$A = \begin{bmatrix} 1 & 2 & 3 \\ 2 & 1 & -1 \\ 0 & 3 & 7 \end{bmatrix}, \quad R = \begin{bmatrix} 1 & 0 & 2 \\ 0 & 1 & -1 \\ 0 & 0 & 0 \end{bmatrix}$$

The basis is the set of transposes of the non-zero rows of the reduced form R of A. Specifically, the basis is $\{X_1, X_2\}$ where

$$X_1 = \begin{bmatrix} 1 \\ 0 \\ 2 \end{bmatrix} X_2 \begin{bmatrix} 0 \\ 1 \\ -1 \end{bmatrix}$$

2.72 (a) Since the nullity of A is $5 - 2 = 3$, it suffices to show that the X_i are independent. For this we reduce either A or A^t where $A = [X_1, X_2, X_2]$ to obtain a rank 3 matrix.

2.78 (a) The first two rows are independent and the last two are linear combinations of them. (The third row is the sum of the first two and the fourth is twice the first plus the second.) Thus the rank is two.

(b) Any pair of independent columns is a basis.

(c) The dimension of the nullspace is $6 - 2 = 4$ so the given vectors do not span the nullspace.

(d) $T = [1, 0, 0, 0, 0, 0]^t$

(e) Any vector of the form $X = T + c_1 X_1 + c_2 X_2$ will work, where T is as in (d) and the X_i are as in (c).

2.80 The rank of A is $n - d$ so the dimension of the nullspace of A^t is $m - (n - d) = m - n + d$.

CHAPTER 3

LINEAR TRANSFORMATIONS

3.1 THE LINEARITY PROPERTIES

Problems begin on page **157**

EXERCISES

3.1 The vertices of the square are $A = [0,0]^t$, $B = [1,0]^t$, $C = [1,1]^t$, and $D = [0,1]^t$. In each case we plot the points $A' = MA = [0,0]^t$, $B' = MB$, $C' = MC$, and $D' = MD$. We then connect A' to B', B' to C', C' to D', and D' to A' with line segments. See Figure 3.1.

3.3 It is easily computed that $A([x,y]^t) = [2x,3y]^t$. Let $u = 2x$ and $v = 3y$. Then $u^2/4 + v^2/9 = (2x)^2/4 + (3y)^2/9 = x^2 + y^2 = 1$ showing that $[u,v]^t$ lies on the ellipse. Conversely, if $[u,v]^t$ is a point on the ellipse then the same equation shows that $[x,y]^t = [u/2,v/3]^t$ lies on the circle, proving that the image of the circle is the ellipse.

Solutions Manual to Accompany Linear Algebra: Ideas and Applications, Fourth Edition. Richard Penney.
© 2016 John Wiley & Sons, Inc. Published 2016 by John Wiley & Sons, Inc.

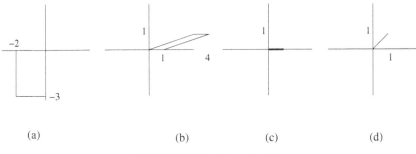

FIGURE 3.1   Exercise 3.1.

3.5 We want either: $A[1,0]' = [2,4]'$ and $A[0,1]' = [2,4]'$ or

$A[1,0]' = [4,2]'$ and $A[0,1]' = [2,4]'$. Hence either $A = \begin{bmatrix} 2 & 4 \\ 4 & 2 \end{bmatrix}$

or $A = \begin{bmatrix} 4 & 2 \\ 2 & 4 \end{bmatrix}$.

3.6 We want either: $A[2,4]' = [1,0]'$ and $A[4,2]' = [0,1]'$ or

$A[2,4]' = [0,1]'$ and $A[4,2]' = [1,0]'$. Consider the first case. Let

$$A = \begin{bmatrix} a & b \\ c & d \end{bmatrix}$$

We need $2a + 4b = 1$, $4a + 2b = 0$, $2c + 4d = 0$, and $4c + 2d = 1$. These equations are easily solved to produce the matrix on the left below. Similarly the second case produces matrix on the right.

$$A = \begin{bmatrix} -1/6 & 1/3 \\ 1/3 & -1/6 \end{bmatrix} \text{ or } A = \begin{bmatrix} 1/3 & -1/6 \\ -1/6 & 1/3 \end{bmatrix}$$

3.8 Let the transformation be denoted T. Then

$$T\left(\begin{bmatrix} x \\ y \end{bmatrix}\right) = \begin{bmatrix} x \\ -y \end{bmatrix} = \begin{bmatrix} 1 & 0 \\ 0 & -1 \end{bmatrix} \begin{bmatrix} x \\ y \end{bmatrix}$$

The 2×2 matrix on the right is the desired matrix.
For the proof, see Figure 3.2.

3.9 (c) When viewed from the positive z axis, the transformation rotates counterclockwise θ radian about the z-axis.

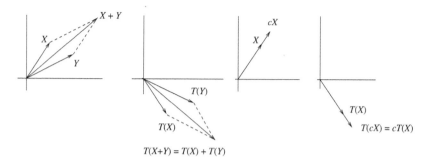

FIGURE 3.2 Exercise 3.8.

3.10 Assume A satisfies $AX_1 = Y_1$ and $AX_2 = Y_2$. Then since $X_2 = 2X_1$,

$$Y_2 = AX_2 = A(2X_1) = 2AX_1 = 2Y_1$$

But $Y_2 = [5, 6]^t \neq 2Y_1$ so there is no such matrix A.

3.12 (a) Note that

$$[2, 1, 1]^t = 2[1, 1, 1]^t - [0, 1, 1]^t$$

Hence

$$
\begin{aligned}
A[2, 1, 1]^t &= A(2[1, 1, 1]^t - [0, 1, 1]^t) \\
&= 2A[1, 1, 1]^t - 2A[0, 1, 1]^t \\
&= 2[1, -2]^t - [1, 1]^t = [1, -5]^t
\end{aligned}
$$

(c) The ith column of A is AI_i where I_i is the ith column of the identity matrix. Hence we need to find AI_i, which requires writing I_i as a linear combination of $X_1 = [2, 1, 1]^t$, $X_2 = [1, 1, 1]^t$, and $X_3 = [0, 0, 1]^t$ for $i = 1, 2, 3$. This in turn requires solving three systems of equations:

$$
\begin{aligned}
a_1 X_1 + a_2 X_2 + a_3 X_3 &= I_1 \\
b_1 X_1 + b_2 X_2 + b_3 X_3 &= I_2 \\
c_1 X_1 + c_2 X_2 + c_3 X_3 &= I_3
\end{aligned}
$$

The augmented matrices of these systems are

$$\begin{bmatrix} 2 & 1 & 0 & 1 \\ 1 & 1 & 0 & 0 \\ 1 & 1 & 1 & 0 \end{bmatrix} \quad \begin{bmatrix} 2 & 1 & 0 & 0 \\ 1 & 1 & 0 & 1 \\ 1 & 1 & 1 & 0 \end{bmatrix} \quad \begin{bmatrix} 2 & 1 & 0 & 0 \\ 1 & 1 & 0 & 0 \\ 1 & 1 & 1 & 1 \end{bmatrix}$$

The corresponding reduced forms are

$$\begin{bmatrix} 1 & 0 & 0 & 1 \\ 0 & 1 & 0 & -1 \\ 0 & 0 & 1 & 0 \end{bmatrix} \quad \begin{bmatrix} 1 & 0 & 0 & -1 \\ 0 & 1 & 0 & 2 \\ 0 & 0 & 1 & -1 \end{bmatrix} \quad \begin{bmatrix} 1 & 0 & 0 & 0 \\ 0 & 1 & 0 & 0 \\ 0 & 0 & 1 & 1 \end{bmatrix}$$

The coefficients of X_i are the solutions of the corresponding systems. We find

$$I_1 = X_1 - X_2$$
$$I_2 = -X_1 + 2X_2 - X_3$$
$$I_3 = X_3$$

The columns A_i of A are then

$$A_1 = AI_1 = AX_1 - AX_2$$
$$A_2 = AI_2 = -AX_1 + 2AX_2 - AX_3$$
$$A_3 = AI_3 = AX_3$$

Hence

$$A = \begin{bmatrix} 0 & -2 & 3 \\ 3 & 0 & -5 \end{bmatrix}$$

3.14 **(a)** Any pair of points will work as long as neither entry is 0.

(b) The the lines $x = n$ are the set of points of the form $[n, y]'$. Since $T([n, y]') = [n, y(1 + n^2)]'$ their image is the line $x = n$. Similarly the image of the line $y = m$ is the set of points of the form $[x, m(1 + x^2)]'$. The graph is the parabola $y = m(1 + x^2)$. The graphs for the desired values of m and n are shown in Figure 3.3.

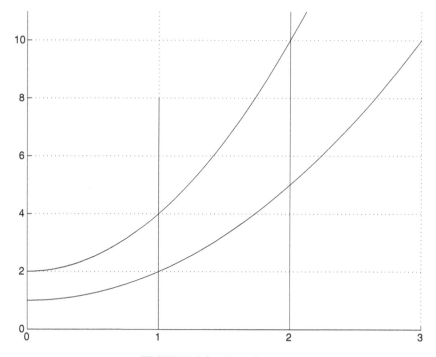

FIGURE 3.3   Exercise 3.14.

3.15 (a) Let $X = [x_1, y_1, z_1]^t$ and $Y = [x_2, y_2, z_2]^t$. Then

$$
\begin{aligned}
T(X) + T(Y) &= [2x_1 + 3y_1 - 7z_1, 0]^t + [2x_2 + 3y_2 - 7z_2, 0]^t \\
&= [2(x_1 + x_2) + 3(y_1 + y_2) - 7(z_1 + z_2), 0]^t \\
&= T(X + Y)
\end{aligned}
$$

Also,

$$
\begin{aligned}
T(c[x, y, z]^t) = T([cx, cy, cz]^t) &= [2cx + 3cy - 7cz, 0]^t \\
&= cT([x, y, z])
\end{aligned}
$$

The matrix that describes T is

$$
\begin{bmatrix} 2 & 3 & -7 \\ 0 & 0 & 0 \end{bmatrix}
$$

because

$$\begin{bmatrix} 2 & 3 & -7 \\ 0 & 0 & 0 \end{bmatrix} \begin{bmatrix} x \\ y \\ z \end{bmatrix} = \begin{bmatrix} 2cx + 3cy - 7cz \\ 0 \end{bmatrix}$$

(c) Let $X = [x_1, y_1]^t$ and $Y = [x_2, y_2]^t$. Then

$$\begin{aligned} T(X) + T(Y) &= [2x_1 + 3y_1, 2y_1 + x_1, -3x_1 - y_1]^t \\ &\quad + [2x_2 + 3y_2, 2y_2 + x_2, -3x_2 - y_2]^t \\ &= [2(x_1 + x_2) + 3(y_1 + y_2), 2(y_1 + y_2) + (x_1 + x_2), \\ &\quad - 3(x_1 + x_2) - (y_1 + y_2)]^t \\ &= T(X + Y) \end{aligned}$$

Also,

$$T(c[x, y]^t) = T([cx, cy]^t) = [2cx + 3cy, 2cy + cx, -3cx - cy]^t$$
$$= cT([x, y, z])$$

The matrix that describes T is

$$\begin{bmatrix} 2 & 3 \\ 1 & 2 \\ -3 & -1 \end{bmatrix}$$

because

$$\begin{bmatrix} 2 & 3 \\ 1 & 2 \\ -3 & -1 \end{bmatrix} \begin{bmatrix} x \\ y \end{bmatrix} = \begin{bmatrix} 2x + 3y \\ 2y + x \\ -3x - y \end{bmatrix}$$

3.16 The image is the x-axis. The plane maps onto 0. The set in question is the plane $2x + 3y - 7z = -3$ which is a plane parallel to the original plane.

3.22 From the given there are non-zero scalars such that

$$aX_1 + bX_2 + cX_3 = 0$$

Hence

$$aT(X_1) + bT(X_2) + cT(X_3) = T(aX_1 + bX_2 + cX_3) = T(0) = 0$$

proving that $a = b = c = 0$.

Alternatively, from the given, for some i, j, k and scalars a, b, $X_i = aX_j + bX_k$ where i, j and k are all distinct. Then

$$T(X_i) = T(aX_j + bX_k) = aT(X_j) + bT(X_k)$$

3.23 The nullspace is non-empty because $T(0) = 0$. (See Exercise 3.21 on page 162.) Let X and Y belong to the nullspace of T. Then for scalars s and t,

$$T(sX + tY) = sT(X) + tT(Y)$$
$$= s0 + t0 = 0$$

proving that $sX + tY$ belongs to the nullspace.

3.24 Let $Y_1 = T(X_1)$ and $Y_2 = T(X_2)$ belong to $T(\mathcal{V})$ and let s and t be scalars. Then

$$sT(Y_1) + tT(Y_2) = T(sY_1 + tY_2)$$

showing that the image is closed under linear combinations.

3.2 MATRIX MULTIPLICATION (COMPOSITION)

Problems begin on page **173**

EXERCISES

3.26

(a) $BC = \begin{bmatrix} 8 & 13 & -3 \\ -2 & -3 & 1 \end{bmatrix}$, $AB = \begin{bmatrix} 2 & 1 \\ 4 & 4 \end{bmatrix}$

$A(BC) = (AB)C = \begin{bmatrix} 4 & 7 & -1 \\ 12 & 20 & -4 \end{bmatrix}.$

3.27

(a)
$$AB = \begin{bmatrix} 2 & 1 \\ 4 & 4 \end{bmatrix}, \quad BA = \begin{bmatrix} 8 & 10 \\ -2 & -2 \end{bmatrix}$$

(c)
$$(AB)^t = \begin{bmatrix} 2 & 4 \\ 1 & 4 \end{bmatrix} = B^t A^t, \quad A^t B^t = \begin{bmatrix} 8 & -2 \\ 10 & -2 \end{bmatrix}$$

3.28 $(ABC)^t = C^t (AB)^t = C^t (B^t A^t) = C^t B^t A^t.$

3.30

(a)
$$U = \begin{bmatrix} 1 & 0 & 1 & 2 \\ 0 & 1 & 2 & 2 \\ 2 & 3 & 0 & 0 \\ 0 & -1 & 0 & 0 \end{bmatrix}, \quad V = \begin{bmatrix} 1 & 2 & 0 & 0 & 0 \\ 2 & 2 & 0 & 0 & 0 \\ 2 & 3 & 1 & 2 & 0 \\ 0 & -1 & 2 & 3 & -1 \end{bmatrix}$$

3.35 One should first answer (b). Then the answer to (a) is found by multiplying the vertices of the unit square by B. Specifically:

(b)

$$B = AR_{\pi/4}$$
$$= \begin{bmatrix} 1 & 1 \\ 0 & 1 \end{bmatrix} \begin{bmatrix} \cos\frac{\pi}{4} & -\sin\frac{\pi}{4} \\ \sin\frac{\pi}{4} & \cos\frac{\pi}{4} \end{bmatrix}$$
$$= \frac{1}{2} \begin{bmatrix} 2\sqrt{2} & 0 \\ \sqrt{2} & \sqrt{2} \end{bmatrix}$$

(a) The quadrilateral with vertices $[0,0]^t$, $[\sqrt{2}, \sqrt{2}/2]^t$, $[\sqrt{2}, \sqrt{2}]^t$, $[0, \sqrt{2}/2]^t$.

3.37 We compute that

$$A^2 = \begin{bmatrix} 1 & 1 \\ 0 & 1 \end{bmatrix} \begin{bmatrix} 1 & 1 \\ 0 & 1 \end{bmatrix} = \begin{bmatrix} 1 & 2 \\ 0 & 1 \end{bmatrix}$$
$$A^3 = A^2 A$$
$$= \begin{bmatrix} 1 & 2 \\ 0 & 1 \end{bmatrix} \begin{bmatrix} 1 & 1 \\ 0 & 1 \end{bmatrix} = \begin{bmatrix} 1 & 3 \\ 0 & 1 \end{bmatrix}$$

We guess that

$$A^n = \begin{bmatrix} 1 & n \\ 0 & 1 \end{bmatrix}$$

To prove it, assume that we have computed that A^n is given by the above formula for some n. Then

$$A^{n+1} = A^n A$$
$$= \begin{bmatrix} 1 & n \\ 0 & 1 \end{bmatrix} \begin{bmatrix} 1 & 1 \\ 0 & 1 \end{bmatrix} = \begin{bmatrix} 1 & n+1 \\ 0 & 1 \end{bmatrix}$$

Hence the formula is true for all n by mathematical induction. The image is the quadrilateral defined by the vectors

$$B = \begin{bmatrix} 1 & n \\ 0 & 1 \end{bmatrix} \begin{bmatrix} 1 \\ 0 \end{bmatrix} = \begin{bmatrix} 1 \\ 0 \end{bmatrix}$$

and

$$C = \begin{bmatrix} 1 & n \\ 0 & 1 \end{bmatrix} \begin{bmatrix} 1 \\ 0 \end{bmatrix} = \begin{bmatrix} 0 \\ 1 \end{bmatrix} = \begin{bmatrix} 0 \\ n \end{bmatrix}$$

3.39 We do (b) before (a).

(b)

$$C = A^2$$
$$= \begin{bmatrix} 2 & 0 \\ 0 & 3 \end{bmatrix} \begin{bmatrix} 2 & 0 \\ 0 & 3 \end{bmatrix}$$
$$= \begin{bmatrix} 4 & 0 \\ 0 & 9 \end{bmatrix}$$

(a) Let

$$\begin{bmatrix} u \\ v \end{bmatrix} = A^2 \begin{bmatrix} x \\ y \end{bmatrix}$$
$$= \begin{bmatrix} 4 & 0 \\ 0 & 9 \end{bmatrix} \begin{bmatrix} x \\ y \end{bmatrix}$$
$$= \begin{bmatrix} 4x \\ 9y \end{bmatrix}$$

Hence

$$x = \frac{u}{4}$$
$$y = \frac{v}{9}$$

Then $[x, y]'$ satisfies

$$x^2 + y^2 = 1$$

if and only if $[u, v]'$ satisfies

$$\frac{u^2}{16} + \frac{v^2}{81} = 1$$

Thus the image is the ellipse defined by this equation.

3.40 We do part (b) first

    (b) From the solution to Exercise 3.8 on page 44 of this manual, the reflection is described by multiplication by

$$T = \begin{bmatrix} 1 & 0 \\ 0 & -1 \end{bmatrix}$$

Hence

$$M = TR_{\pi/4} = \begin{bmatrix} \sqrt{2}/2 & -\sqrt{2}/2 \\ -\sqrt{2}/2 & -\sqrt{2}/2 \end{bmatrix}$$

$$Y = R_{\pi/4}T = \begin{bmatrix} \sqrt{2}/2 & \sqrt{2}/2 \\ \sqrt{2}/2 & -\sqrt{2}/2 \end{bmatrix}$$

    (a) The images are obtained by multiplying the vertices of the square by the matrices M and Y in part (b). Thus my answer is the square with vertices M_0, M_1, M_2, and M_3 where $M_0 = M0 = 0$ and

$$M_1 = \begin{bmatrix} \sqrt{2}/2 & -\sqrt{2}/2 \\ -\sqrt{2}/2 & -\sqrt{2}/2 \end{bmatrix} \begin{bmatrix} 1 \\ 0 \end{bmatrix} = \begin{bmatrix} \sqrt{2}/2 \\ -\sqrt{2}/2 \end{bmatrix}$$

$$M_2 = \begin{bmatrix} \sqrt{2}/2 & -\sqrt{2}/2 \\ -\sqrt{2}/2 & -\sqrt{2}/2 \end{bmatrix} \begin{bmatrix} 0 \\ 1 \end{bmatrix} = \begin{bmatrix} -\sqrt{2}/2 \\ -\sqrt{2}/2 \end{bmatrix}$$

$$M_3 = \begin{bmatrix} \sqrt{2}/2 & -\sqrt{2}/2 \\ -\sqrt{2}/2 & -\sqrt{2}/2 \end{bmatrix} \begin{bmatrix} 1 \\ 1 \end{bmatrix} = \begin{bmatrix} 0 \\ -\sqrt{2} \end{bmatrix}$$

Your answer is the square with vertices Y_0, Y_1, Y_2, and Y_3 where $Y_0 = Y0 = 0$ and

$$Y_1 = \begin{bmatrix} \sqrt{2}/2 & \sqrt{2}/2 \\ \sqrt{2}/2 & -\sqrt{2}/2 \end{bmatrix} \begin{bmatrix} 1 \\ 0 \end{bmatrix} = \begin{bmatrix} \sqrt{2}/2 \\ \sqrt{2}/2 \end{bmatrix}$$

$$Y_2 = \begin{bmatrix} \sqrt{2}/2 & \sqrt{2}/2 \\ \sqrt{2}/2 & -\sqrt{2}/2 \end{bmatrix} \begin{bmatrix} 0 \\ 1 \end{bmatrix} = \begin{bmatrix} \sqrt{2}/2 \\ -\sqrt{2}/2 \end{bmatrix}$$

$$Y_3 = \begin{bmatrix} \sqrt{2}/2 & \sqrt{2}/2 \\ \sqrt{2}/2 & -\sqrt{2}/2 \end{bmatrix} \begin{bmatrix} 1 \\ 1 \end{bmatrix} = \begin{bmatrix} \sqrt{2} \\ 0 \end{bmatrix}$$

See Figure 3.1.

3.41 Let $B = [B_1, B_2]$ where the B_i are columns. Since

$$AB = [AB_1, AB_2]$$

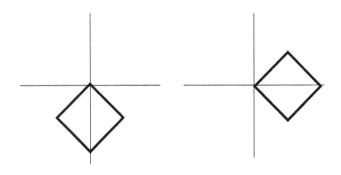

My answer   Your answer

FIGURE 3.1   Exercise 3.37.

$AB = 0$ if and only if both B_1 and B_2 belong to the nullspace of A. Finding the nullspace of A which requires solving the system $AX = 0$. The coefficient matrix and reduced form for this system are respectively

$$\begin{bmatrix} 1 & 2 & 1 & 0 \\ 1 & 1 & 1 & 0 \end{bmatrix}, R = \begin{bmatrix} 1 & 0 & 1 & 0 \\ 0 & 1 & 0 & 0 \end{bmatrix}$$

Thus the third variable is a free variable and the nullspace is spanned by $X = [-1, 0, 1]^t$. The columns of B must be multiples of X. For example:

$$B = \begin{bmatrix} -1 & -2 \\ 0 & 0 \\ 1 & 2 \end{bmatrix}$$

Since each column is a multiple of a single column, if $A \neq 0$, A will have rank 1.

3.45 You could let A be arbitrary and $B = I$, or $B = 0$, or $B = A$, or $B = A^2, \ldots$

3.47 Almost any randomly chosen A and B will work. The equality is true if and only if

$$(A + B)(A - B) = A^2 - B^2$$
$$BA - AB = 0$$
$$BA = AB$$

3.48 Let $a \neq 0$. The following matrices work:

$$A = \begin{bmatrix} 0 & a \\ 0 & 0 \end{bmatrix}, B = \begin{bmatrix} 0 & 0 \\ a & 0 \end{bmatrix}$$

3.51 Any matrix of the form $A = \begin{bmatrix} \pm 1 & 0 & 0 \\ 0 & \pm 1 & 0 \\ 0 & 0 & \pm 1 \end{bmatrix}$ works. There are eight such matrices.

3.59 Let s, t be scalars and $X, Y \in \mathcal{V}$. Then

$$
\begin{aligned}
S \circ T(sX + tY) &= S(T(sX + tY)) \\
&= S(sT(X) + tT(Y)) \\
&= sS(T(X)) + tS(T(Y)) \\
&= sS \circ T(X) + tS \circ T(Y)
\end{aligned}
$$

3.2.2 Applications to Graph Theory II

Problems begin on page **181**

SELF-STUDY QUESTIONS

3.1 One can take either of the two flights from Indianapolis to Chicago and either of the two return flights for a total of four possible two step connections.

3.2 One can take either of the two flights from Indianapolis to Chicago and then catch the Denver flight for a total of two possible two step connections.

3.3 The desired matrix is the square of the given matrix which is

$$
\begin{bmatrix} 5 & 1 & 4 \\ 6 & 4 & 1 \\ 1 & 6 & 5 \end{bmatrix}
$$

3.3 INVERSES

Problems begin on page **190**

EXERCISES

3.63 **(a)** One reduces the matrix on the left obtaining the martix on the right.

$$
\begin{bmatrix} 1 & 1 & 1 & y_1 \\ 1 & 2 & 2 & y_2 \\ 2 & 3 & 4 & y_3 \end{bmatrix}
\quad
\begin{bmatrix} 1 & 0 & 0 & -y_2 + 2y_1 \\ 0 & 1 & 0 & -y_3 + 2y_2 \\ 0 & 0 & 1 & y_3 - y_1 - y_2 \end{bmatrix}
$$

Since

$$\begin{bmatrix} -y_2 + 2y_1 \\ -y_3 + 2y_2 \\ y_3 - y_1 - y_2 \end{bmatrix} = \begin{bmatrix} 2 & -1 & 0 \\ 0 & 2 & -1 \\ -1 & -1 & 1 \end{bmatrix} \begin{bmatrix} y_1 \\ y_2 \\ y_3 \end{bmatrix}$$

our answer is the 3×3 matrix on the right.

3.64 In each case, we form the double matrix $[A|I]$ and reduce until either the left matrix has a row of zeros, in which case the matrix is not invertible, or we obtain a matrix of the form $[I|B]$, in which case our answer is B. We find that (c) and (e) are not invertible. The inverses of the others are:

(a) $\dfrac{1}{10}\begin{bmatrix} -26 & 15 & -12 \\ 20 & -10 & 10 \\ 12 & -5 & 4 \end{bmatrix}$ (b) $\dfrac{1}{2}\begin{bmatrix} -1 & 1 & 1 \\ 1 & -1 & 1 \\ 1 & 1 & -1 \end{bmatrix}$

(d) $\dfrac{1}{4}\begin{bmatrix} 2 & -1 & 1 \\ -2 & 5 & -1 \\ 0 & -2 & 2 \end{bmatrix}$

(g) $\dfrac{1}{3}\begin{bmatrix} 23 & 17 & -17 & -13 \\ 2 & 2 & -2 & -1 \\ 9 & 9 & -6 & -6 \\ -15 & -12 & 12 & 9 \end{bmatrix}$ (h) $\dfrac{1}{12}\begin{bmatrix} 12 & -6 & 0 & 0 \\ 0 & 6 & -4 & 0 \\ 0 & 0 & 4 & -3 \\ 0 & 0 & 0 & 3 \end{bmatrix}$

(i) $\begin{bmatrix} a^{-1} & 0 & 0 & 0 \\ 0 & b^{-1} & 0 & 0 \\ 0 & 0 & c^{-1} & 0 \\ 0 & 0 & 0 & d^{-1} \end{bmatrix}$

3.65 In each case, we compute $A^{-1}Y$ where A^{-1} is as given in the solution to Exercise 3.64 given above. The answers are

(a) $\frac{1}{5}[-16, 15, -7]^t$,

(b) $[2, 1, 0]^t$,

(c) not invertible,

(d) $\frac{1}{4}[3,5,2]'$,

(e) not invertible,

(f) $\frac{1}{4}[22,8,-11,4]'$,

(g) $\frac{1}{3}[-46,-4,-15,33]'$,

(h) $[0,0,0,1]'$,

(i) $[\frac{1}{a},\frac{2}{b},\frac{3}{c},\frac{4}{d}]'$,

(j) $\frac{1}{3}[-1,-2,4,8]'$,

(k) $\frac{1}{2}[-1,4,1,4]'$,

(l) $[-1,0,3]'$,

(m) $[0,1,2]'$,

(n) $[-1,0,3]'$

(o) $\frac{1}{3}[-6,9,4]'$,

(p) $[1,0,1]'$.

3.66 Part (c) is most easily done by inspection.

(a) $A_2 = 2A_1, A_3 = 3A_1, B_2 = 2B_1, B_3 = 3B_1$.

3.68 Make one row (or column) a linear combination of other rows (resp. columns). It will not be invertible because the rank will not be 5.

3.69 We compute the inverse of the coefficient matrix and multiply it with the vector of constants obtaining the result shown below. This was more work because finding the inverse requited reducing the 3×6 double matrix $[A|I]$ while solving the system directly only required reducing a 3×4 matrix.

$$X = \begin{bmatrix} \frac{50}{9} & 5/9 & -\frac{17}{9} \\ -\frac{11}{9} & -2/9 & 5/9 \\ -\frac{8}{9} & 1/9 & 2/9 \end{bmatrix} \begin{bmatrix} 1 \\ 2 \\ 7 \end{bmatrix} = \begin{bmatrix} -\frac{59}{9} \\ \frac{20}{9} \\ \frac{8}{9} \end{bmatrix}$$

3.71 We reduce the double matrix

$$\begin{bmatrix} a & b & | & 1 & 0 \\ c & d & | & 0 & 1 \end{bmatrix}$$

Note that either $a \neq 0$ or $b \neq 0$ since otherwise $ad - bc = 0$. Suppose first that $a \neq 0$.

Then the steps in the row reduction are

$$\left[\begin{array}{cc|cc} a & b & 1 & 0 \\ c & d & 0 & 1 \end{array}\right] \rightarrow \left[\begin{array}{cc|cc} 1 & \frac{b}{a} & \frac{1}{a} & 0 \\ c & d & 0 & 1 \end{array}\right] \rightarrow$$

$$\left[\begin{array}{cc|cc} 1 & \frac{b}{a} & \frac{1}{a} & 0 \\ 0 & \frac{da-bc}{a} & -\frac{c}{a} & 1 \end{array}\right] \rightarrow \left[\begin{array}{cc|cc} 1 & \frac{b}{a} & \frac{1}{a} & 0 \\ 0 & 1 & -\frac{c}{da-bc} & \frac{a}{da-bc} \end{array}\right]$$

$$\rightarrow \left[\begin{array}{cc|cc} 1 & 0 & \frac{d}{ad-bc} & -\frac{b}{ad-bc} \\ 0 & 1 & -\frac{c}{ad-bc} & \frac{a}{ad-bc} \end{array}\right]$$

Hence

$$A^{-1} = (ad - bc)^{-1} \begin{bmatrix} d & -b \\ -c & a \end{bmatrix}$$

If $a = 0$, then $c \neq 0$ and we interchange the rows in the double matrix and proceed analogously to the above computation. Ultimately we get the same formula for A^{-1}.

3.72 We reduce the double matrix $[A|I]$ obtaining $\begin{bmatrix} 1 & -a & ac - b \\ 0 & 1 & -c \\ 0 & 0 & 1 \end{bmatrix}$

3.74 (b) Proof:

$$(I - N)(I + N + N^2) = I + N + N^2 - N - N^2 - N^3$$
$$= I - N^3 = I$$

It follows from Theorem 3.10 on page 188 that

$$(I - N)^{-1} = I + N + N^2$$

3.76 $$X = BC = \begin{bmatrix} 3 & 4 & 7 \\ 2 & 2 & 4 \\ 6 & 8 & 14 \end{bmatrix}$$

3.77 $I = -A^2 - 3A = A(-A - 3I)$. The result follows from Theorem 3.10 on page 188.

3.80 If $ABX = Y$ then $BX = A^{-1}Y$ so $X = B^{-1}(A^{-1}Y) = (B^{-1}A^{-1})Y$. Hence $(AB)^{-1} = B^{-1}A^{-1}$ since the matrix that describes a linear transformation is unique.

3.84 $(A^{-1})^t A^t = (AA^{-1})^t = I^t = I$

3.86 In general $(Q^{-1}DQ)^n = Q^{-1}D^nQ$. For the proof we note first that

$$(Q^{-1}DQ)^2 = (Q^{-1}DQ)(Q^{-1}DQ) = Q^{-1}D^2Q$$

Assume that we have proved that for some n $(Q^{-1}DQ)^n = Q^{-1}D^nQ$. Then

$$(Q^{-1}D^nQ)(Q^{-1}DQ) = Q^{-1}D^nDQ = Q^{-1}D^{n+1}Q$$

The claim follows by induction.

3.88 (a) We solve systems $AX = I_j$ for $j = 1, 2$. The corresponding augmented matrices and their reduced forms are respectively

$$\begin{bmatrix} 1 & 2 & 1 & 1 \\ 1 & 1 & 1 & 0 \end{bmatrix}, \begin{bmatrix} 1 & 0 & 1 & -1 \\ 0 & 1 & 0 & 1 \end{bmatrix}$$

$$\begin{bmatrix} 1 & 2 & 1 & 0 \\ 1 & 1 & 1 & 1 \end{bmatrix}, \begin{bmatrix} 1 & 0 & 1 & 2 \\ 0 & 1 & 0 & -1 \end{bmatrix}$$

The general solutions to these systems are respectively

$$X_1 = [-1, 1, 0]^t + s[-1, 0, 1]$$
$$X_2 = [2, -1, 0]^t + t[-1, 0, 1]$$

Then $B = [X_1, X_2]$. For example if we choose $s = t = 0$ we find

$$B = \begin{bmatrix} -1 & 2 \\ 1 & -1 \\ 0 & 0 \end{bmatrix}$$

(c) $2 = \text{rank}(I) = \text{rank}(AB) \le \text{rank}(B) \le 2$.

3.91 Let B be the left inverse of A. If $AX = 0$, then $0 = BAX = X$, showing that the nullspace is zero.

Alternatively if $BA = I$ then $n = \text{rank}(I) = \text{rank}(BA) \leq \text{rank}(A) \leq n$. Hence rank $A = n$. It follows from the rank-nullity theorem that $A = 0$.

3.92 **(b)** The rank of a 3×2 matrix is at most 2; hence the rows of A must be dependent. Now, let the rows of A be A_i and those of C be C_i. Then, $A_3 = 3A_2 - 6A_1$. Hence

$$C_3 = A_3 B = 3A_2 B - 6A_1 B = 3C_2 - 6C_1$$

Thus the rank of C is at most 2 showing that C is not invertible.

3.3.2 Applications to Economics

Problems begin on page **202**

SELF-STUDY QUESTIONS

3.4 Reasoning as in in Example 3.9 we find

$$\begin{bmatrix} c \\ e \end{bmatrix} \approx \begin{bmatrix} 1.1481 & 0.7407 \\ 0.2469 & 10.2346 \end{bmatrix} \begin{bmatrix} 4000 \\ 3000 \end{bmatrix} \approx \begin{bmatrix} 6800 \\ 4700 \end{bmatrix}$$

3.5 **(a)** The industrial demand vector is the product of the output matrix times the consumption matrix. Hence in this case the industrial demand is

$$\begin{bmatrix} 0.2 & 0.1 & 0.3 \\ 0.1 & 0.3 & 0.2 \\ 0.2 & 0.2 & 0.2 \end{bmatrix} \begin{bmatrix} 20 \\ 15 \\ 10 \end{bmatrix} = \begin{bmatrix} 8.5 \\ 8.5 \\ 9.0 \end{bmatrix}$$

In particular the demand for manufacturing is 8.5 units.

(b) From the discussion in Example 3.10 the level of production is $X = (I - C)^{-1}D$ which on this case is

$$\begin{bmatrix} 1.4607 & 0.3933 & 0.6461 \\ 0.3371 & 1.6292 & 0.5337 \\ 0.4494 & 0.5056 & 1.5449 \end{bmatrix} \begin{bmatrix} 50 \\ 35 \\ 20 \end{bmatrix} \approx \begin{bmatrix} 100.0 \\ 84.6 \\ 71.0 \end{bmatrix}$$

(c) The entries of $(I - C)^{-1}$ are all positive.

3.4 THE *LU* FACTORIZATION

Problems begin on page **212**

EXERCISES

3.100 Z and X are respectively:

$$(a) \begin{bmatrix} 6 \\ -1 \\ -8 \\ 6 \end{bmatrix} \begin{bmatrix} -8 \\ 3 \\ -7 \\ 6 \end{bmatrix} \quad (b) \begin{bmatrix} 3 \\ -4 \\ 3 \end{bmatrix} \begin{bmatrix} 5 \\ -7 - t \\ 3 - t \\ t \end{bmatrix}$$

3.101

(a)
$$L = \begin{bmatrix} 1 & 0 & 0 \\ 2 & 1 & 0 \\ 1 & \frac{1}{2} & 1 \end{bmatrix}, \quad U = \begin{bmatrix} 1 & 4 & -1 \\ 0 & 2 & 4 \\ 0 & 0 & 1 \end{bmatrix}$$

(d)
$$L = \begin{bmatrix} 1 & 0 & 0 \\ \frac{1}{2} & 1 & 0 \\ \frac{1}{2} & 0 & 1 \end{bmatrix}, \quad U = \begin{bmatrix} 4 & 2 & 2 \\ 0 & 1 & 1 \\ 0 & 0 & 0 \end{bmatrix}$$

3.102 This follows from the first paragraph in the proof of Theorem 3.12 on page 206.

3.103 If $[U|B]$ is produced by reducing $[A|I]$, then $BA = U$. This follows from the first paragraph in the proof of Theorem 3.12 on page 206.

3.104 Let U be as in formula (3.30) on page 207 and let D be a diagonal matrix having the same diagonal entries as U. Then $\tilde{U} = D^{-1}U$ is unipotent and satisfies $A = LD\tilde{U}$ as desired.

$$D = \begin{bmatrix} 1 & 0 & 0 \\ 0 & -3 & 0 \\ 0 & 0 & \frac{2}{3} \end{bmatrix}, \quad \tilde{U} = \begin{bmatrix} 1 & 2 & 1 \\ 0 & 1 & \frac{1}{3} \\ 0 & 0 & 1 \end{bmatrix}$$

3.106
$$P = PI = \begin{bmatrix} 0 & 1 & 0 \\ 1 & 0 & 0 \\ 0 & 0 & 1 \end{bmatrix}$$

3.110 You must interchange rows 2 and 3 of A for the LU factorization to exist. Let P be the corresponding permutation matrix. The LU decomposition for PA is

$$P = \begin{bmatrix} 1 & 0 & 0 \\ 0 & 0 & 1 \\ 0 & 1 & 0 \end{bmatrix}, \quad PA = \begin{bmatrix} 1 & 0 & 0 \\ 2 & 1 & 0 \\ 2 & 0 & 1 \end{bmatrix}\begin{bmatrix} 1 & 1 & 3 \\ 0 & -1 & -2 \\ 0 & 0 & -1 \end{bmatrix}$$

Since $P^2 = I$, $A = PLU$.

3.5 THE MATRIX OF A LINEAR TRANSFORMATION

Problems begin on page **230**

EXERCISES

3.114 Let $X = [1, 1]^t$ and $Y = [1, -1]^t$. Then $[x, y]^t = x'X + y'Y = [x' + y', x' - y']^t$ so

$$1 = xy = (x' + y')(x' - y') = (x')^2 - (y')^2$$

which is of the desired form.

3.117 Below, P the point matrix, $C = P^{-1}$ is the coordinate matrix, and $X = C[1, 2, 3]^t$ is the coordinate vector for $[1, 2, 3]^t$.

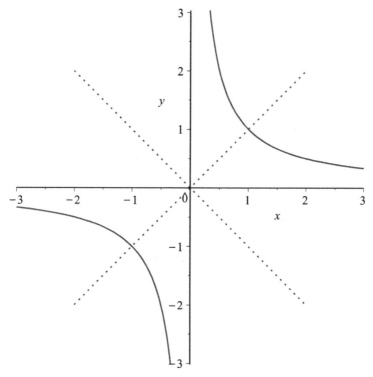

FIGURE 3.2 Exercise 3.114

(b)

$$P = \begin{bmatrix} 1 & 2 & 1 \\ -2 & 3 & 1 \\ 1 & 2 & 0 \end{bmatrix} \quad C = \begin{bmatrix} 2/7 & -2/7 & 1/7 \\ -1/7 & 1/7 & 3/7 \\ 1 & 0 & -1 \end{bmatrix} \quad X = \begin{bmatrix} 1/7 \\ 10/7 \\ -2 \end{bmatrix}$$

3.118 Let the polynomials be respectively p_i, $i = 1, 2, \ldots$ and the corresponding coordiante vectors be X_i. In each exercise the X_i satisfy the same dependency equations as the p_i. Call the polynomials p_i.

    (b) $X_1 = [1, -4, 4, 4]^t, X_2 = [2, -1, 2, 1]^t, X_3 = [-17, -, -8, 2]^t$,
    $p_3 = -p_1 + 2p_2$.

3.119 In each case let the elements of $\mathcal{B}$ be denoted $X_j\, j = 1, \ldots, n$ where $n = 2$ or 3. Then $T_A(X_j) = \lambda_j X_j$. Hence the coordinate vector for $T(X_j)$ is $\lambda_j I_j$ where I_j is the jth standard basis element for $\mathbb{R}^n$. It follows from the comments immediately preceding Example 3.22

on page 227 that the matrix of T_A is $M = [\lambda_1 I_1, \ldots, \lambda_n I_n]$. Explicitly,
our answers are: (a) $\begin{bmatrix} 5 & 0 \\ 0 & 15 \end{bmatrix}$, (c) $\begin{bmatrix} 1 & 0 & 0 \\ 0 & 2 & 0 \\ 0 & 0 & -3 \end{bmatrix}$

3.120 To find X we solve the system $AX = 0$. Since in each case rank $A = 2$, the nullity is one and X is unique up to scalar multiples. We give an acceptable choice of X below. For the proof, note that in each case $A[1, 0, 0]' = A_1$, $A[0, 1, 0]' = A_2$, and $AX = 0$. Hence $(A[1, 0, 0]')' = [1, 0]'$, $(A[0, 1, 0]')' = [0, 1]'$, and $(AX)' = [0, 0]'$. The claim follows from formula (3.40) on page 227.

Possible value of X: (b) $[0, -1, 1]$.

3.121 The point matrix for the standard basis for $\mathbb{R}^2$ is the 2×2 identity matrix I. The coordinate matrix is $C = I^{-1} = I$. Hence from formula (3.36) on page 220 $M = AP_B$. Specifically

$$P = \begin{bmatrix} 1 & 1 & 1 \\ 1 & 2 & 0 \\ 1 & -1 & -1 \end{bmatrix}$$

Hence:

(a) $M = AP = \begin{bmatrix} 4 & -11 & -5 \\ 6 & -5 & -3 \end{bmatrix}$

3.126 The calculations are similar to those done in Example 3.21 on page 226. Specifically, we let

$$X = \begin{bmatrix} a & b \\ c & d \end{bmatrix}$$

We then compute

$$L(X) = \begin{bmatrix} u & v \\ w & z \end{bmatrix}$$

using the given formula for L. Then in the standard basis for $M(2, 2)$, $(L(X))' = [u, v, w, z]'$. By inspection we find a matrix M such that

$[u, v, w, z]^t = M[a, b, c, d]^t$. In part (a), for example,

$$(L(X))' = (AX)' = \left(\begin{bmatrix} a + 2c & b + 2d \\ 3a + 4c & 3b + 4d \end{bmatrix} \right)'$$
$$= [a + 2c, 3a + 4c, b + 2d, 3b + 4d]^t$$

This equals $M[a, b, c, d]^t$ where M is given in (a) below. The answers to the other parts follow.

(a) $\begin{bmatrix} 1 & 0 & 2 & 0 \\ 0 & 1 & 0 & 2 \\ 3 & 0 & 4 & 0 \\ 0 & 3 & 0 & 4 \end{bmatrix}$, (c) $\begin{bmatrix} 1 & 2 & 2 & 4 \\ 3 & 4 & 6 & 8 \\ 3 & 6 & 4 & 8 \\ 9 & 12 & 12 & 16 \end{bmatrix}$, (e) $\begin{bmatrix} 2 & 0 & 0 & 0 \\ 0 & 1 & 1 & 0 \\ 0 & 1 & 1 & 0 \\ 0 & 0 & 0 & 2 \end{bmatrix}$

3.130 The calculations are similar to those done in Example 3.20 on page 224. Specifically, we compute

$$L(a + bx + cx^2) = a' + b'x + c'x^2$$

using the given formula for L. By inspection we find a matrix M such that $[a', b', c']^t = M[a, b, c]^t$. In part (a), for example, $L(y) = b + 2cx$ which has coordinates $[b, 2c, 0]^t = M[a, b, c]^t$ where M is given in (a) below. The answers to the other parts follow.

(a) $\begin{bmatrix} 0 & 1 & 0 \\ 0 & 0 & 2 \\ 0 & 0 & 0 \end{bmatrix}$, (c) $\begin{bmatrix} -1 & 1 & 0 \\ 0 & -1 & 2 \\ 0 & 0 & -1 \end{bmatrix}$, (e) $\begin{bmatrix} -7 & 0 & 0 \\ 0 & -4 & 0 \\ 0 & 0 & 1 \end{bmatrix}$.

3.133 For part (a) we find

$$L(1) = 1' = 0 = 0 + 0(x - 1) + 0(x - 1)^2$$
$$L(x) = x' = 1 = 1 + 0(x - 1) + 0(x - 1)^2$$
$$L(x^2) = (x^2)' = 2x$$
$$= 2((x - 1) + 1) = 2 + 2(x - 1) + 0(x - 1)^2$$

According to formula (3.40) on page 227 the the coefficients on the right side of the preceding expressions form the coefficients of M. Our answer is as in part (a). The other parts are solved similarly.

$$(a) \begin{bmatrix} 0 & 1 & -2 \\ 0 & 0 & 2 \\ 0 & 0 & 0 \end{bmatrix}, (c) \begin{bmatrix} -1 & 2 & -3 \\ 0 & -1 & 4 \\ 0 & 0 & -1 \end{bmatrix}, (e) \begin{bmatrix} -7 & 7 & -7 \\ 0 & -4 & 8 \\ 0 & 0 & 1 \end{bmatrix}.$$

CHAPTER 4

DETERMINANTS

4.1 DEFINITION OF THE DETERMINANT

Problems begin on page **249**

EXERCISES

4.1 **(a)** $1 \cdot 2 - (-3) \cdot 4 = 14.$

(c) 0

$$2 \begin{vmatrix} 1 & 1 \\ 2 & 2 \end{vmatrix} - 0 \begin{vmatrix} -1 & 1 \\ -2 & 2 \end{vmatrix} + 1 \begin{vmatrix} -1 & 1 \\ -2 & 2 \end{vmatrix} = 2 \cdot 0 - 0 \cdot 0 + 1 \cdot 0 = 0$$

(e) $-16,$

(g) $0,$

$$\begin{vmatrix} 1 & 1 \\ 2 & 1 \end{vmatrix} - 0 \begin{vmatrix} 2 & 1 \\ 3 & 1 \end{vmatrix} + \begin{vmatrix} 2 & 1 \\ 3 & 2 \end{vmatrix} = -1 - 0 + 1 = 0$$

Solutions Manual to Accompany Linear Algebra: Ideas and Applications, Fourth Edition. Richard Penney.
© 2016 John Wiley & Sons, Inc. Published 2016 by John Wiley & Sons, Inc.

(i) 34, (The 3×3 determinants are evaluated as in parts (c)–(h).)

$$2\begin{vmatrix} 1 & 1 & 1 \\ 0 & 3 & 2 \\ 0 & 0 & 5 \end{vmatrix} - 0\begin{vmatrix} 1 & 1 & 1 \\ 0 & 3 & 2 \\ 1 & 0 & 5 \end{vmatrix} + 2\begin{vmatrix} 1 & 1 & 1 \\ 0 & 0 & 2 \\ 1 & 0 & 5 \end{vmatrix} - 0\begin{vmatrix} 1 & 1 & 1 \\ 0 & 0 & 3 \\ 1 & 0 & 0 \end{vmatrix}$$
$$= 2 \cdot 15 - 0 + 2 \cdot 2 - 0 = 34$$

4.2 Let R_i denote the ith row and C_i the ith column.

   (a) The determinant is not 0.

   (b) $R_3 = 2R_2, C_3 = \frac{1}{2}C_1 + \frac{3}{2}C_2$.

   (c) The determinant is not 0.

   (d) $R_3 = 2R_2 - R_1, C_3 = C_1 + C_2$.

   (e) The determinant is not 0.

4.3

$$\det A = -a_{21}(a_{12}a_{33} - a_{13}a_{32})$$
$$+ a_{22}(a_{11}a_{33} - a_{13}a_{31}) - a_{23}(a_{11}a_{32} - a_{12}a_{31})$$

4.5 Let A, B, C, D, and E be the 4×4 determinants from the solution to Exercise 4.4 in the order listed above. Then

$$\beta = 3A - 12B + 6C - 6D - 9E$$
$$\gamma = 2A - 5B + 3C - 3D - 4E$$
$$\delta = \ \ A - 7B + 3C - 3D + 13E$$

from which $\beta = \gamma + \delta$ is clear. This demonstrates the row additivity.

4.7 **(b)** Expand along the row not assumed equal to the others. The resulting 2×2 determinants are zero by (a), (c) Similar to (b).

4.11 **(a)** The 2×2 case is a direct computation. For the general case, assume by induction that the result is true for all $k \times k$ matrices A' and B' where $k < n$. Let A and B be $n \times n$ matrices. Then From formula (4.3) on page 240

$$\det(A - \lambda B) = (a_{11} - \lambda b_{11}) \det(A_{11} - \lambda B_{11})$$
$$- (a_{12} - \lambda b_{12}) \det(A_{12} - \lambda B_{12}) + \cdots$$
$$+ (-1)^{n-1}(a_{1n} - \lambda b_{1n}) \det(A_{1n} - \lambda B_{1n})$$

The result follows from the observation that by the inductive hypothesis each of the terms $\det(A_{1j} - \lambda B_{1j})$ is a polynomial of degree at most $n - 1$.

(b) From part (a) we may write $P_A(\lambda) = a_n \lambda^n + a_{n-1} \lambda^{n-1} + \ldots a_0$. Hence

$$\lim_{n \to \infty} \lambda^{-n} P_A(\lambda) = a_n.$$

On the other hand

$$\lim_{n \to \infty} \lambda^{-n} \det(A - \lambda I) = \lim_{n \to \infty} \det(\lambda^{-1} A - I) = \det(-I).$$

Part (b) follows.

4.2 REDUCTION AND DETERMINANTS

Problems begin on page **258**

EXERCISES

4.12 (a)

$$
\begin{vmatrix} 1 & 0 & 1 \\ 2 & 1 & 1 \\ 3 & 2 & 1 \end{vmatrix}
=
\begin{vmatrix} 1 & 0 & 1 \\ 0 & 1 & -1 \\ 0 & 2 & -2 \end{vmatrix}
=
\begin{vmatrix} 1 & 0 & 1 \\ 0 & 1 & -1 \\ 0 & 0 & 0 \end{vmatrix}
= 0
$$

(c)

$$
\begin{vmatrix} 2 & 0 & 2 & 0 \\ 1 & 1 & 1 & 1 \\ 0 & 0 & 3 & 2 \\ 1 & 0 & 0 & 5 \end{vmatrix}
= -
\begin{vmatrix} 1 & 1 & 1 & 1 \\ 2 & 0 & 2 & 0 \\ 0 & 0 & 3 & 2 \\ 1 & 0 & 0 & 5 \end{vmatrix}
= -
\begin{vmatrix} 1 & 1 & 1 & 1 \\ 0 & -2 & 0 & -2 \\ 0 & 0 & 3 & 2 \\ 0 & -1 & -1 & 4 \end{vmatrix}
$$

$$
= -
\begin{vmatrix} 1 & 1 & 1 & 1 \\ 0 & -1 & -1 & 4 \\ 0 & 0 & 3 & 2 \\ 0 & -2 & 0 & -2 \end{vmatrix}
= -
\begin{vmatrix} 1 & 1 & 1 & 1 \\ 0 & -1 & -1 & 4 \\ 0 & 0 & 3 & 2 \\ 0 & 0 & 2 & -10 \end{vmatrix}
$$

$$
= -2 \begin{vmatrix} 1 & 1 & 1 & 1 \\ 0 & -1 & -1 & 4 \\ 0 & 0 & 3 & 2 \\ 0 & 0 & 1 & -5 \end{vmatrix} = 2 \begin{vmatrix} 1 & 1 & 1 & 1 \\ 0 & -1 & -1 & 4 \\ 0 & 0 & 1 & -5 \\ 0 & 0 & 3 & 2 \end{vmatrix}
$$

$$
= 2 \begin{vmatrix} 1 & 1 & 1 & 1 \\ 0 & -1 & -1 & 4 \\ 0 & 0 & 1 & -5 \\ 0 & 0 & 0 & 17 \end{vmatrix} = 34
$$

(e)

$$
\begin{vmatrix} 1 & 2 & 3 & 4 \\ 5 & 6 & 7 & 8 \\ 9 & 10 & 11 & 12 \\ 13 & 14 & 15 & 16 \end{vmatrix} = \begin{vmatrix} 1 & 2 & 3 & 4 \\ 0 & -4 & -8 & -12 \\ 0 & -8 & -16 & -24 \\ 0 & -12 & -24 & -36 \end{vmatrix}
$$

$$
= \begin{vmatrix} 1 & 2 & 3 & 4 \\ 0 & -4 & -8 & -12 \\ 0 & 0 & 0 & 0 \\ 0 & 0 & 0 & 0 \end{vmatrix} = 0
$$

4.13 (a)

$$
\begin{vmatrix} 1 & 0 & 1 \\ 2 & 1 & 1 \\ 3 & 2 & 1 \end{vmatrix} = \begin{vmatrix} 1 & 0 & 0 \\ 2 & 1 & -1 \\ 3 & 2 & -2 \end{vmatrix} = \begin{vmatrix} 1 & 0 & 0 \\ 2 & 1 & 0 \\ 3 & 2 & 0 \end{vmatrix} = 0
$$

(c)

$$
\begin{vmatrix} 2 & 0 & 2 & 0 \\ 1 & 1 & 1 & 1 \\ 0 & 0 & 3 & 2 \\ 1 & 0 & 0 & 5 \end{vmatrix} = \begin{vmatrix} 2 & 0 & 0 & 0 \\ 1 & 1 & 0 & 1 \\ 0 & 0 & 3 & 2 \\ 1 & 0 & -1 & 5 \end{vmatrix}
$$

$$
= \begin{vmatrix} 2 & 0 & 0 & 0 \\ 1 & 1 & 0 & 0 \\ 0 & 0 & 3 & 2 \\ 1 & 0 & -1 & 5 \end{vmatrix} = \begin{vmatrix} 2 & 0 & 0 & 0 \\ 1 & 1 & 0 & 1 \\ 0 & 0 & 3 & 0 \\ 1 & 0 & -1 & \frac{17}{3} \end{vmatrix} = 34
$$

(e)

$$
\begin{vmatrix} 1 & 2 & 3 & 4 \\ 5 & 6 & 7 & 8 \\ 9 & 10 & 11 & 12 \\ 13 & 14 & 15 & 16 \end{vmatrix} = \begin{vmatrix} 1 & 0 & 0 & 0 \\ 5 & -4 & -8 & -12 \\ 9 & -8 & -16 & -24 \\ 13 & -12 & -24 & -36 \end{vmatrix}
$$

$$
= \begin{vmatrix} 1 & 0 & 0 & 0 \\ 5 & -4 & 0 & 0 \\ 9 & -8 & 0 & 0 \\ 13 & -12 & 0 & 0 \end{vmatrix} = 0
$$

4.14 (a)

$$
\begin{vmatrix} 1 & 4 \\ -3 & 2 \end{vmatrix} = \begin{vmatrix} 1 & 4 \\ 0 & 14 \end{vmatrix} = 14
$$

(e)

$$
\begin{vmatrix} 0 & 5 & 1 \\ -1 & 1 & 3 \\ -2 & -2 & 2 \end{vmatrix} = \begin{vmatrix} -1 & 1 & 3 \\ 0 & -4 & -4 \\ 0 & 5 & 1 \end{vmatrix} = 4 \begin{vmatrix} -1 & 1 & 3 \\ 0 & 1 & 1 \\ 0 & 0 & -4 \end{vmatrix} = -16
$$

(i)

$$
\begin{vmatrix} 2 & 0 & 2 & 0 \\ 1 & 1 & 1 & 1 \\ 0 & 0 & 3 & 2 \\ 1 & 0 & 0 & 5 \end{vmatrix} = - \begin{vmatrix} 1 & 0 & 0 & 5 \\ 1 & 1 & 1 & 1 \\ 0 & 0 & 3 & 2 \\ 2 & 0 & 2 & 0 \end{vmatrix}
$$

$$
= - \begin{vmatrix} 1 & 0 & 0 & 5 \\ 0 & 1 & 1 & -4 \\ 0 & 0 & 3 & 2 \\ 0 & 0 & 2 & -10 \end{vmatrix} = - \begin{vmatrix} 1 & 0 & 0 & 5 \\ 0 & 1 & 1 & -4 \\ 0 & 0 & 3 & 2 \\ 0 & 0 & 0 & -\frac{34}{3} \end{vmatrix} = 34
$$

4.15 (a)

$$
\begin{vmatrix}
2a & 2b & 2c \\
3d - a & 3e - b & 3f - c \\
4g + 3a & 4h + 3b & 4i + 3c
\end{vmatrix}
= 2
\begin{vmatrix}
a & b & c \\
3d & 3e & 3f \\
4g & 4h & 4i
\end{vmatrix}
$$
$$
= 2 \cdot 3 \cdot 4 \cdot 5 = 120
$$

(b)

$$
\begin{vmatrix}
a + 2d & b + 2e & c + 2f \\
g & h & i \\
d & e & f
\end{vmatrix}
=
\begin{vmatrix}
a & b & c \\
g & h & i \\
d & e & f
\end{vmatrix}
= -
\begin{vmatrix}
a & b & c \\
d & e & f \\
g & h & i
\end{vmatrix}
= -5.
$$

4.18 For some $k = 1, 2$ or 3 and scalars c and d, $A_k = cA_i + dA_j$ where $\{i, j, k\} = \{1, 2, 3\}$. From the row exchange property we may assume that $i = 1$, $j = 2$, and $k = 3$. The argument is then the same as that in Exercise 4.17.

4.20 See the solution to Exercise 4.18.

4.25

$$
\det(QBQ^{-1}) = \det(Q)\det(B)\det(Q^{-1})
$$
$$
= \det(Q)\det(Q^{-1})\det(B) = \det(I)\det(B) = \det B
$$

4.2.1 Volume

Problems begin on page **263**

EXERCISES

4.28 The figures are shown in Figure 4.1. The answers are

(a)

$$
\det
\begin{bmatrix}
2 & 0 \\
0 & 1
\end{bmatrix}
= 2
$$

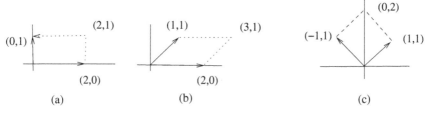

FIGURE 4.1   Exercise 4.28

(b)

$$\det \begin{bmatrix} 2 & 0 \\ 1 & 1 \end{bmatrix} = 2$$

(c)

$$\det \begin{bmatrix} 1 & -1 \\ 1 & 1 \end{bmatrix} = 2$$

4.29   Let V be the volume.

(a)

$$\begin{vmatrix} 1 & 2 & 1 \\ 1 & -1 & 0 \\ 1 & 1 & 1 \end{vmatrix} = -1, V = 1$$

(b)

$$\begin{vmatrix} 2 & 1 & 0 \\ -1 & 2 & 0 \\ 0 & 0 & 1 \end{vmatrix} = 5, V = 5$$

(c)

$$\begin{vmatrix} 2 & 0 & 2 \\ -5 & 4 & 0 \\ 1 & 1 & 1 \end{vmatrix} = -10, V = 10$$

4.3 A FORMULA FOR INVERSES

Problems begin on page **268**

EXERCISES

4.34 From Cramer's rule

$$x = \frac{a_x}{a} = -\frac{9}{4}, \; y = \frac{a_y}{a} = -\frac{5}{4}$$

$$z = \frac{a_z}{a} = \frac{8}{4}, \; w = \frac{a_w}{a} = \frac{15}{4}$$

$$a = \begin{vmatrix} 1 & 2 & 1 & 1 \\ 1 & 0 & 3 & -1 \\ 2 & 1 & 1 & 1 \\ 1 & -1 & 1 & 0 \end{vmatrix}, \quad a_x = \begin{vmatrix} 1 & 2 & 1 & 1 \\ 0 & 0 & 3 & -1 \\ 0 & 1 & 1 & 1 \\ 1 & -1 & 1 & 0 \end{vmatrix}$$

$$a_y = \begin{vmatrix} 1 & 1 & 1 & 1 \\ 1 & 0 & 3 & -1 \\ 2 & 0 & 1 & 1 \\ 1 & 1 & 1 & 0 \end{vmatrix}, \quad a_z = \begin{vmatrix} 1 & 2 & 1 & 1 \\ 1 & 0 & 0 & -1 \\ 2 & 1 & 0 & 1 \\ 1 & -1 & 1 & 0 \end{vmatrix},$$

$$a_w = \begin{vmatrix} 1 & 2 & 1 & 1 \\ 1 & 0 & 3 & 0 \\ 2 & 1 & 1 & 0 \\ 1 & -1 & 1 & 1 \end{vmatrix}$$

4.37 $c_1(x) = e^{-x}(1 + x^2)\cos x - e^{-x}x\sin x, c_2(x) = x\cos x + (1 + x^2)\sin x$.

4.38 $A_{21}^{-1} = (-1)^3 a_{12}/\det A$ where
   (a) Not invertible.
   (c) $-a_{12} = - \begin{vmatrix} 0 & 2 & 0 \\ 0 & 3 & 2 \\ 0 & 0 & 5 \end{vmatrix} = 0, \det(A) = 32$
   (e) Not invertible.

4.39   Let $B = A^{-1}$. Then

$$b_{13} = (-1)^{1+3}(\det A_{31})/(\det A)$$

$$= \begin{vmatrix} 2 & 1 & 1 \\ 0 & 3 & -1 \\ -1 & 1 & 0 \end{vmatrix} /6 = 4/6$$

4.40   From the solution to Example 4.10 on page 265, $A^{-1} = B$ where

$$B = \begin{bmatrix} 0 & -1/2 & 1 \\ 1 & 3/2 & -4 \\ -1 & -1/2 & 3 \end{bmatrix}$$

From Cramer's rule

$$B_{32}^{-1} = (-1)^{1+2}\frac{|B_{23}|}{|B|}$$

$$= -\frac{\begin{vmatrix} 0 & -1/2 \\ -1 & -1/2 \end{vmatrix}}{1/2}$$

$$= 1$$

You get $A_{32} = 1$ because $(A^{-1})^{-1} = A$.

CHAPTER 5

EIGENVECTORS AND EIGENVALUES

5.1 EIGENVECTORS

Problems begin on page **279**

EXERCISES

5.1 In each part, compare X with AX to see if $AX = \lambda X$.

  (a) X is an eigenvector with eigenvalue 0, Y is not, Z is not. (By definition, eigenvectors must be nonzero.)

  (b) X is an eigenvector with eigenvalue 3. Y is not an eigenvector.

  (c) Both X and Y are eigenvectors with eigenvalues 0 and 3, respectively.

5.2 To compute the characteristic polynomials, we compute $\det(A - \lambda I)$ by expanding along the first row. We then solve the system $(A - \lambda I)X = 0$.

Solutions Manual to Accompany Linear Algebra: Ideas and Applications, Fourth Edition. Richard Penney.
© 2016 John Wiley & Sons, Inc. Published 2016 by John Wiley & Sons, Inc.

(a) $p(\lambda) = -\lambda^2(\lambda - 3)$, eigenvalues and corresponding basis: $\lambda = 0$: $[1, 0, -1]^t$ and $[1, -1, 0]^t$, $\lambda = 3$, $[1, 1, 1]^t$,

$$A^n B = 5[3^{n-1}, 3^{n-1}, 3^{n-1}]^t, \ n \geq 1$$

5.3 We obtain the eigenvalue by comparing AZ with Z for each given eigenvector Z.

(a) $2, 2, 4, B = -\frac{1}{4}X + \frac{3}{4}Z, A^n B = 2^{n-2}X + 3(4^{n-1})Z$.

5.4 To check your answer compute AY to see that $AY = \lambda Y$.

(a) Any vector $Y = aX + bY$ where $a < 0$ and $b > -2a$.

5.5 An $n \times n$ matrix is diagonalizable if and only if it has n linearly independent eigenvectors. In each case we compute and factor $P_A(\lambda) = \det(A - \lambda I)$. For each root λ of $P_A(\lambda)$ we solve the system $(A - \lambda I)X = 0$, finding the spanning vectors. The set of all vectors so produced is guaranteed to be independent. If there are n of them A will be diagonalizable. We provide detailed solutions to parts (a) and (e)

(a) $\lambda = -3$, $[2, 1]^t$; $\lambda = 2$, $[3/2, 1]^t$, diagonalizable, $p_A(\lambda) = (\lambda + 3)(\lambda - 2)$.
Solution:
For each eigenvalue λ, we give $B = A - \lambda I$, the reduced form R of B augmented by a column of 0's from which one obtains the stated spanning vectors.

$$A + 3I = \begin{bmatrix} -15 & 30 \\ -10 & 20 \end{bmatrix}, R = \begin{bmatrix} 1 & -2 & 0 \\ 0 & 0 & 0 \end{bmatrix}$$

$$A - 2I = \begin{bmatrix} -20 & 30 \\ -10 & 15 \end{bmatrix}, R = \begin{bmatrix} 1 & -3/2 & 0 \\ 0 & 0 & 0 \end{bmatrix}, X = \begin{bmatrix} 3/2 \\ 1 \end{bmatrix}$$

(b) $\lambda = -3$, $[7, 3]^t$; $\lambda = 2$, $[3, 2]^t$, diagonalizable, $P_A(\lambda) = (\lambda + 3)(\lambda - 2)$.

(c) $\lambda = 2$, $[-2, 1, 0]^t$, $[1, 0, 1]^t$; $\lambda = 3$, $[0, 2/5, 1]^t$, diagonalizable, $P_A(\lambda) = (\lambda - 2)^2(\lambda - 3)$.

Solution:
For each eigenvalue λ, we give $B = A - \lambda I$, the reduced form R of B augmented by a column of 0's from which one obtains the stated spanning vectors.

$$A - 2I = \begin{bmatrix} 0 & 0 & 0 \\ -2 & -4 & 2 \\ -5 & -10 & 5 \end{bmatrix}, \begin{bmatrix} 1 & 2 & -1 & 0 \\ 0 & 0 & 0 & 0 \\ 0 & 0 & 0 & 0 \end{bmatrix}$$

$$A - 3I = \begin{bmatrix} -1 & 0 & 0 \\ -2 & -5 & 2 \\ -5 & -10 & 4 \end{bmatrix}, \begin{bmatrix} 1 & 0 & 0 & 0 \\ 0 & 1 & -2/5 & 0 \\ 0 & 0 & 0 & 0 \end{bmatrix}$$

(d) $\lambda = 1, [1, 2/3, 1]^t$, deficient over the real numbers (but not over the complex numbers), $P_A(\lambda) = -(\lambda - 1)(\lambda^2 + 2)$.

(e) $\lambda = 2, [1, 1/2, 1]^t, \lambda = 3, [1, 0, 1]^t, p_A(\lambda) = -(\lambda - 3)(\lambda - 2)^2$, deficient.

5.6 We obtain the eigenvalues either by consulting the answers to Exercise 5.3 or by comparing AZ with Z for each given eigenvector Z. For each eigenvalue λ we solve the system $(A - \lambda I)X = 0$, finding the spanning vectors.

(b) $\lambda = 2$, basis $\{X' = [0, 1, 1]^t, Y' = [1/2, 1, 0]^t\}$; $\lambda = 4$, basis $\{Z' = [1, 1, 1]^t\}$. This basis is consistent with the basis $\{X, Z, Y\}$ from Exercise 5.3.b because $X = 4Y' - 2X')$ and $Y = 2Y'$. Since the span of $\{X', Y'\}$ is two dimensional and $\{X, Y\}$ is linearly independent it follows that $\{X, Y\}$ and $\{X', Y'\}$ are both bases for the $\lambda = 2$ eigenspace.

5.7 The $\lambda = 2$ eigenspace is the $x - y$ plane and the $\lambda = 3$ eigenspace is the z axis.

5.8 The $\lambda = 0$ eigenspace is the null space.

5.11 Since λ is an eigenvector, there is an X such that $AX = \lambda X$. Multiplying both sides of this equation by A, we see that

$$A(AX) = A(\lambda X) = \lambda AX$$
$$A^2 X = \lambda(\lambda X)$$
$$IX = (\lambda^2)X$$
$$X = \lambda^2 X$$

Since X is non-zero, last equation can only be true if $\lambda^2 = 1$. To see this, let x_i be a non-zero entry for X. The above equation implies that $x_i = \lambda^2 x_i$. Dividing by x_i proves that $\lambda^2 = 1$ so $\lambda = \pm 1$.

5.13   Suppose $AX = \lambda X$. Then

$$A^{-1}(AX) = \lambda A^{-1}X$$
$$\lambda^{-1}X = A^{-1}X$$

5.14   *Hint:* See Exercise 5.8 and Theorem 5.2 on page 278.

5.1.2   Application to Markov Processes

Problems begin on page **285**

5.17   We find a spanning vector for the $\lambda = 1$ eigenspace and divide by the sum of its entries. We obtain (to 4 decimal accuracy):

(a) $\frac{1}{12}[5,7]^t = [0.4167, 0.5833]^t$

(b) $\frac{1}{41}[9,17,15]^t = [0.2195, 0.4146, 0.3658]^t$

5.18   (a)   To two decimal accuracy the equilibrium vector is $X = [5,7]^t/12 = [0.42, 0.58]^t$. Also $P^4[1,0]^t = [0.42, 0.58]^t$.

(b)   To two decimal accuracy the equilibrium vector is $X = [1,4]^t/5 = [0.20, 0.80]^t$. Also $P^8[1,0]^t = [0.20, 0.80]^t$.

5.21   (a)   Tuesday: $P[1,1,1]^t = [0.4667, 0.3333, 0.2]^t$,   Wednesday: $P[0.4667, 0.3333, 0.2]^t = [0.4933, 0.3067, 0.2]^t$.

(b)   To find the equilibrium distribution find a spanning vector for the $\lambda = 1$ eigenspace and divide by the sum of its entries to make the sum of the entries equal to one. We get $X = \frac{1}{25}[13,7,5]^t$.

(c) $Y_1 = [3,2,-5]^t$, $Y_2 = [-1,1,0]^t$.

(d)   To four decimal places

$$V_0 = X - \frac{2}{75}Y_1 + \frac{8}{75}Y_2$$
$$A^5 V_0 = X - \frac{2}{75}(0.5)^5 Y_2$$
$$= [0.5200, 0.2800, 0.2000]^t$$

(e) Note that the entries in both Y_1 and Y_2 sum to 0 while the entries in X and V_0 sum to one. Hence equation (5.8) implies $1 = a + c_1 01 + c_2 0$. Hence, for $n > 1$

$$V_0 = X + c_1 Y_1 + c_2 Y_2$$
$$A^n V_0 = A^n X + c_1 A^n Y_1 + c_2 A^n Y_2$$
$$= X + c_2 (0.5)^n Y_2 \to X$$

5.22 (a) The transition matrix is

$$P = \begin{bmatrix} .6 & .15 & .10 \\ .2 & .7 & .10 \\ .2 & .15 & .8 \end{bmatrix}$$

5.2 DIAGONALIZATION

Problems begin on page **290**

EXERCISES

5.27 The columns of Q are the basis elements from each eigenspace and the diagonal entries of D are the corresponding eigenvalues.

(a)

$$Q = \begin{bmatrix} 1 & 1 & 0 \\ 1 & 0 & 1 \\ -1 & 1 & 1 \end{bmatrix}, \quad D = \begin{bmatrix} 6 & 0 & 0 \\ 0 & -3 & 0 \\ 0 & 0 & 0 \end{bmatrix},$$

$$A^n = Q \begin{bmatrix} 6^n & 0 & 0 \\ 0 & (-3)^n & 0 \\ 0 & 0 & 0 \end{bmatrix} Q^{-1}$$

(b)

$$Q = \begin{bmatrix} 1 & 0 & 1 \\ 0 & 1 & 1 \\ -2 & 1 & 1 \end{bmatrix}, \quad D = \begin{bmatrix} 2 & 0 & 0 \\ 0 & 2 & 0 \\ 0 & 0 & 4 \end{bmatrix},$$

$$A^n = Q \begin{bmatrix} 2^n & 0 & 0 \\ 0 & 2^n & 0 \\ 0 & 0 & 4^n \end{bmatrix} Q^{-1}$$

5.29 We find the eigenvalues and eigenvectors using the process described in the solution to Exercise 5.5 described on page 77 of this manual.

(a) Not diagonalizable over $\mathbb{R}$.

(b) Not diagonalizable.

(c)

$$Q = \begin{bmatrix} 1 & 1 & 0 & 0 \\ 1 & -1 & 0 & 0 \\ 0 & 0 & 2 & -1 \\ 0 & 0 & -1 & 1 \end{bmatrix}, \quad D = \begin{bmatrix} 4 & 0 & 0 & 0 \\ 0 & -2 & 0 & 0 \\ 0 & 0 & -2 & 0 \\ 0 & 0 & 0 & -3 \end{bmatrix}$$

(d)

$$Q = \begin{bmatrix} 1 & 1 \\ 3 & 2 \end{bmatrix}, \quad D = \begin{bmatrix} 2 & 0 \\ 0 & -1 \end{bmatrix}$$

(f)

$$Q = \begin{bmatrix} 0 & 1 & 2 \\ 1 & 0 & 1 \\ 1 & 0 & 0 \end{bmatrix}, \quad D = \begin{bmatrix} 3 & 0 & 0 \\ 0 & 1 & 0 \\ 0 & 0 & 2 \end{bmatrix}$$

5.33 From Theorem 5.4 on page 287, $A = QDQ^{-1}$ where D is a diagonal matrix with diagonal entries ± 1. Then $D^2 = I$ and $A^2 = QD^2Q^{-1} = QIQ^{-1} = I$.

5.35 From Theorem 5.4 on page 287, $A = QDQ^{-1}$ where D is a diagonal matrix with diagonal entries 2 or 4. Since both 2 and 4 are roots of the equation $x^2 - 6x + 8 = 0$, we see $D^2 - 6D + 8I = 0$. Then

$$A^2 - 6A + 8I = QD^2Q^{-1} - 6QDQ^{-1} + 8QQ^{-1}$$
$$= Q(D^2 - 6D + 8I)Q^{-1} = 0$$

5.37 The $\lambda = -5$ eigenspace is guaranteed to be one dimensional. Hence, to be diagonalizable the $\lambda = 2$ eigenspace must be two dimensional and $A - 2I$ must have rank 1. Since

$$A - 2I = \begin{bmatrix} 0 & a & b \\ 0 & -7 & c \\ 0 & 0 & 0 \end{bmatrix}$$

it follows that A is diagonalizable if and only if $[a, b] = d[-7, c]$ for some scalar d.

5.2.1 Application to Systems of Differential Equations

Problems begin on Page **295**

SELF-STUDY QUESTIONS

5.1 Since $x = -C_1 e^{2t} + C_2 e^{4t}$ and $y = C_1 e^{2t} + C_2 e^{4t}$

$$x' = -2C_1 e^{2t} + 4C_2 e^{4t}$$
$$y' = 2C_1 e^{2t} + 4C_2 e^{4t}$$
$$\begin{bmatrix} 3 & 1 \\ 1 & 3 \end{bmatrix} \begin{bmatrix} x \\ y \end{bmatrix} = \begin{bmatrix} -2C_1 e^{2t} + 4C_2 e^{4t} \\ 2C_1 e^{2t} + 4C_2 e^{4t} \end{bmatrix} = \begin{bmatrix} x' \\ y' \end{bmatrix}$$

as desired.

5.2
$$A = \begin{bmatrix} 3 & -1 & 1 \\ 1 & 2 & 0 \\ 2 & -1 & -1 \end{bmatrix}$$

5.3 **(a)** The eigenvalues are $6, -3, 0$. Hence the general solution is $W = C_1 e^{6t} X + C_2 e^{-3t} Y + C_3 Z$ where X, Y, and Z are as stated in Exercise 5.3.a on page 280.

(b) The eigenvalues are $2, 2, 4$. Hence the general solution is $W = C_1 e^{2t} X + C_2 e^{2t} Y + C_3 e^{4t} Z$ where X, Y, and Z are as stated in Exercise 5.3.b on page 280.

5.3 COMPLEX EIGENVECTORS

Problems begin on page **304**

EXERCISES

5.44 $AB = \begin{bmatrix} -1 + 3i & -5 + 3i \\ -3 + 4i & -6 \end{bmatrix}$, $BA = \begin{bmatrix} 7 - i & 2 + 9i \\ 1 + 11i & -14 + 4i \end{bmatrix}$

5.46 Expansion of $\det(A - \lambda I)$ along the first row shows that

$$p_A(\lambda) = \lambda^3 - 6\lambda^2 - 42\lambda + 343$$

Since -7 is an eigenvalue, $\lambda + 7$ is a factor of $p_A(\lambda)$. Trial and error, or long division of polynomials, shows that

$$p_A(\lambda) = (\lambda + 7)\left(\lambda^2 - 13\lambda + 49\right)$$

Hence we obtain the roots $-7, 6.5 \pm (3\sqrt{3}/2)i$ which are the eigenvalues.

5.48 $M = \begin{bmatrix} a & -b \\ b & a \end{bmatrix}$. Eigenvalues: $a \pm bi$.

5.50 With $z = a + bi$, $w = c + di$, compute $\overline{zw}$ and $\overline{z}\,\overline{w}$ to show $\overline{zw} = \overline{z}\,\overline{w}$. Then use formula (3.4) on page 167.

5.51 The set B' spans $\mathbb{C}^2$ over $\mathbb{R}$ since

$$[a + ib, c + id]^t = a[1, 0]^t + b[i, 0]^t + c[0, 1]^t + d[0, i]^t$$

Also, if the right side of this formula equals 0, then $a + ib = c + id = 0$ which implies $a = b = c = d = 0$, proving independence.

5.52 Let $z = [z_1, \ldots, z_n]^t \in \mathbb{C}^n$ where $z_j = a_j + ib_j$, $1 \leq j \leq n$. Then

$$z = z_1 I_1 + z_2 I_2 + \cdots + z_n I_n$$
$$= a_1 I_1 + b_1 (iI_1) + \cdots + a_n I_1 + b_n (iI_n)$$

Hence the set $B' = \{I_1, iI_1, \ldots, I_n, iI_n\}$ spans $\mathbb{C}^n$ over $\mathbb{R}$. The independence follows as in Exercise 5.51. Hence the dimension over $\mathbb{R}$ is $2n$.

5.53 Let $B = \{X_j \mid 1 \le j \le n\}$ be a complex basis for $\mathcal{V}$ and let $z \in \mathcal{V}$. Then then for some $z_j = a_j + ib_j \in \mathbb{C}$, $1 \le j \le n$

$$z = z_1 X_1 + z_2 X_2 + \cdots + z_n X_n$$
$$= a_1 X_1 + b_1(iX_1) + \cdots + a_n X_n + b_n(iX_n)$$

Hence the set $B' = \{X_1, iX_1, \ldots, X_n, iX_n\}$ spans C^n over $\mathbb{R}$. To prove independence assume that the right side of the second equation in formula (5.1) equals 0. It follows that right side of the first equation in formula (5.1) equals 0. Hence, from the independence of the X_j, $z_j = 0$ for all j and thus $a_j = b_j = 0$ for all j.

5.55 See the argument after equation (5.24) on page 303.

CHAPTER 6

ORTHOGONALITY

6.1 THE SCALAR PRODUCT IN $\mathbb{R}^n$

Problems begin on page **316**

EXERCISES

6.1 We compute the angle using $\theta = \arccos((X \cdot Y)/(|X|\,|Y|))$.

    (a) $|X - Y| = \sqrt{(3 - (-1))^2 + (4 - 2)^2} = \sqrt{20}$,
$|X| = \sqrt{3^2 + 4^2} = 5$, $|Y| = \sqrt{(-1)^2 + 2^2} = \sqrt{5}$,
$X \cdot Y = 3 \cdot (-1) + 4 \cdot 2 = 5$, $\theta = \arccos(5/(5\sqrt{5})) = 1.1071$
radians.

    (c) $|X - Y| = \sqrt{(1 - (-1))^2 + (2 - 1)^2 + (3 - 2)^2} = \sqrt{6}$,
$|X| = \sqrt{1^2 + 2^2 + 3^2} = \sqrt{14}$, $|Y| = \sqrt{(-1)^2 + 1^2 + 2^2} = \sqrt{6}$,
$X \cdot Y = 1 \cdot (-1) + 2 \cdot 1 + 3 \cdot 2 = 7$, $\theta = \arccos$
$(7/(\sqrt{14}\sqrt{6})) = 0.7017$ radians.

Solutions Manual to Accompany Linear Algebra: Ideas and Applications, Fourth Edition. Richard Penney.
© 2016 John Wiley & Sons, Inc. Published 2016 by John Wiley & Sons, Inc.

6.2 X is perpendicular to Y if and only if $X \cdot Y = 0$. Answers:
(a) Not perpendicular, (c) perpendicular, (d) not perpendicular.

6.5 From formula (6.8), $B \cdot A = |B| \sqrt{3} \cos \theta \geq \frac{3}{2}|B|$ if and only if $\cos \theta \geq \frac{\sqrt{3}}{2}$. This describes the cone consisting of all vectors which make an angle of 30 or less with respect to the vector A.

6.6 (a) $A = [-1, 1, 0]^t$, $B = [-1, -1, -1]^t$, $C = [2, -1, 0]^t$ work
   (b) $[2, 3, -6]^t \cdot (A - B) = [2, 3, -6]^t \cdot [0, 2, 1]^t = 0$ and $[2, 3, -6]^t \cdot (C - B) = [2, 3, -6]^t \cdot [3, 0, 1]^t = 0$.
   (c) $[x, y, z]^t - B = [x + 1, y + 1, z + 1]^t$ which is perpendicular to $[2, 3, 6]^t$ if and only if $0 = [x + 1, y + 1, z + 1]^t \cdot [2, 3, -6]^t = 2x + 3y - 6z - 1$ so the set of solutions is the plane through B perpendicular to $[2, 3, -6]^t$.

6.7 (a) The third basis
   (b) $X' = [\frac{13}{14}, -\frac{2}{3}, -\frac{5}{42}]^t$. See (c) for work.
   (c) From formula (6.12) on page 314

$$X = x_1' P_1 + x_2' P_2 + x_3 P_3 \text{ where}$$
$$x_i' = \frac{X \cdot P_i}{P_i \cdot P_i}$$

Hence

$$
\begin{bmatrix} x_1' \\ x_2' \\ x_3' \end{bmatrix} =
\begin{bmatrix} \frac{1}{14}x_1 - \frac{1}{3}x_2 - \frac{1}{3}x_3 \\ \frac{3}{14}x_1 + \frac{1}{3}x_2 + \frac{1}{3}x_3 \\ \frac{1}{7}x_1 - \frac{1}{3}x_2 - \frac{1}{3}x_3 \end{bmatrix}
$$

$$
= \begin{bmatrix} \frac{1}{14} & -\frac{1}{3} & -\frac{1}{3} \\ \frac{3}{14} & \frac{1}{3} & \frac{1}{3} \\ \frac{1}{7} & -\frac{1}{3} & -\frac{1}{3} \end{bmatrix} \begin{bmatrix} x_1 \\ x_2 \\ x_3 \end{bmatrix}
$$

The above 3×3 matrix is C_B.

6.9 $X' = [\frac{4}{7}, \frac{16}{35}, 0, -\frac{3}{5}]^t$

6.10 Let $P_4 = X$. Then $0 = P_1 \cdot X = P_2 \cdot X = P_3 \cdot X$ is a rank 3 system of equations for X with augmented matrix and reduced form respectively:

$$\begin{bmatrix} 1 & 1 & 1 & 1 & 0 \\ 1 & -2 & 1 & 0 & 0 \\ 1 & 1 & 1 & -3 & 0 \end{bmatrix} \qquad \begin{bmatrix} 1 & 0 & 1 & 0 & 0 \\ 0 & 1 & 0 & 0 & 0 \\ 0 & 0 & 0 & 1 & 0 \end{bmatrix}$$

The general solution is $X = s[-1, 0, 1, 0]^t$.

6.11 Let $P_4 = X$. Then $0 = P_1 \cdot X = P_2 \cdot X = P_3 \cdot X$ is a rank 3 system of equations for X which will have a one dimensional solution set. We may choose P_4 to be any element of this set. It is unique up to scalar multiples. In $\mathbb{R}^n$ the corresponding system will have rank 3 and the solution set is an $(n - 3)$-dimensional subspace. We may choose P_4 to be any non-zero element of this subspace.

6.2 PROJECTIONS: THE GRAM-SCHMIDT PROCESS

Problems begin on page **328**

EXERCISES

6.19 Let P_i, $i = 1, 2, 3$, be the elements of the basis. From the commutativity of the dot product, for the orthogonality we need only check that $P_i \cdot P_j = 0$ for $1 \le i < j \le 3$. We use formula (6.20) on page 321 to find the projections.

(a) Projection: $\frac{1}{5}[8, 4, 4, -12]^t$.

6.20 (b) Let W be the span of $\mathcal{B}_1$. Clearly dim $W = 2$. Also $[3, 1, 4]^t$ and $[-4, 16, -1]^t$ belong to W since

$$[3, 1, 4]^t = -2[-1, 1, -1]^t + [1, 3, 2]^t$$
$$[-4, 16, -1]^t = 7[-1, 1, -1]^t + 3[1, 3, 2]^t$$

(From Theorem 2.5 on page 106 these coefficients may be read off of the row reduced echelon form S of the matrix A having the elements of B_1 as its first two columns and the elements of B_2 as its second two columns.) S and A are as shown below

$$A = \begin{bmatrix} -1 & 1 & 3 & -4 \\ 1 & 3 & 1 & 16 \\ -1 & 2 & 4 & -1 \end{bmatrix} \quad S = \begin{bmatrix} 1 & 0 & -2 & 7 \\ 0 & 1 & 1 & 3 \\ 0 & 0 & 0 & 0 \end{bmatrix}$$

This shows that B_2 spans $\mathcal{W}$ since this set is clearly independent.

(c) and (d): $[\frac{47}{42}, \frac{85}{42}, \frac{40}{21}]^t$. We get the same answer because from Theorem 6.9 on page 324 the projection of a given vector onto a given subspace is unique.

6.21 (e) Let $X = [x, y, z, w]^t$. Then

$$\text{Proj}_{\mathcal{W}}X = \begin{bmatrix} 3x + 6y + 3z - 3w \\ 6x + 19y - z + w \\ 3x - y + 10z - 10w \\ -3x + y - 10z + 10w \end{bmatrix}$$

6.22 (a) $\{[0, 1, 1]^t, [1, 0, 0]^t\}$

   (b) $\{[1, 2, 1, 1]^t, [-2, 1, 1, -1]^t, \frac{1}{7}[-6, -2, 0, 10]^t\}$

6.25 In this exercise, we first compute the Gram-Schmidt basis $\{P_1, \ldots, P_n\}$ for the set $\{A_1, \ldots, A_n\}$ where the A_i are the columns of A. To express the A_i as linear combinations of the P_i in a computer aided hand computation, rather than following the process described in the text it is easier to use Theorem 2.5 on page 106. Explicitly, one computes the reduced echelon form S of the matrix $B = [P_1, \ldots, P_n, A_1, \ldots, A_n]$. The first n columns of B will be the pivot columns, allowing the A_i to be expressed as linear combinations of the P_i. The coefficients may be read off of S. Then Q, D and U are given by formulas (6.28) on page 325 and (6.29) on page 325.

(a)

$$Q = \frac{1}{\sqrt{3}} \begin{bmatrix} 1 & 0 \\ 1 & 1 \\ 0 & 1 \\ 1 & -1 \end{bmatrix}, \quad U = \begin{bmatrix} 1 & 1 \\ 0 & 1 \end{bmatrix}, \quad D = \begin{bmatrix} \sqrt{3} & 0 \\ 0 & \sqrt{3} \end{bmatrix}$$

6.29 (a) We find a basis for the nullspace by row reducing A augmented by a column of zeros. We clear fractions from the basis elements by multiplying by appropriate scalars. We then apply the Gram-Schmidt process to produce an orthogonal basis. We give the reduced form of A, followed by the two basis elements for the null space and the two orthogonal basis elements.

$$\begin{bmatrix} 1 & 3 & 0 & 1/2 & 0 \\ 0 & 0 & 1 & -3/2 & 0 \\ 0 & 0 & 0 & 0 & 0 \end{bmatrix}, \begin{bmatrix} -1 \\ 0 \\ 3 \\ 2 \end{bmatrix}, \begin{bmatrix} -3 \\ 1 \\ 0 \\ 0 \end{bmatrix}, \begin{bmatrix} -1 \\ 0 \\ 3 \\ 2 \end{bmatrix}, \frac{1}{14}\begin{bmatrix} -39 \\ 14 \\ -9 \\ -6 \end{bmatrix}$$

(b) We use formula (6.20) on page 321 applied to the orthogonal basis from part (a) obtaining $[\frac{57}{131}, \frac{171}{131}, -\frac{7}{131}, \frac{39}{131}]^t$.

6.30 (a) $\{[-1, 0, 3, 2]^t, [-3, 1, 0, 0]^t\}$.

6.31 Basis: $\{[1, 3, 1, -1]^t, [5, 15, -19, 31]^t\}$

6.33 The coefficient matrix has the X_i as rows. Hence, if the X_i are independent, the rank is k. From the rank nullity theorem, the dimension of the nullspace (which is $S^\perp$) is $n - k$.

6.3 FOURIER SERIES: SCALAR PRODUCT SPACES

Problems begin on page **341**

EXERCISES

6.40 This follows from the observation that $h(x) = f(x)g(x)$ is odd.

6.41 We compute

$$c_k = (f, q_k) = \int_{-1}^{1} x^3 \sin(k\pi x)\, dx$$
$$= [12/(k^3\pi^3) - 2/(k\pi)](-1)^k$$

Hence $f_o(x) = \sum_{k=1}^{n}[12/(k^3\pi^3) - 2/(k\pi)](-1)^k \sin k\pi x$.

6.43 **(a)** For $n \neq \pm m$

$$\int_{-1}^{1} \cos(n\pi x) \cos(m\pi x)dx$$

$$= 1/2 \left(\frac{\sin((n\pi - m\pi)x)}{n\pi - m\pi} + \frac{\sin((n\pi + m\pi)x)}{n\pi + m\pi} \right) \Big|_{-1}^{1} = 0$$

(b) For $n \neq 0$, (p_n, p_n) equals

$$\int_{-1}^{1} (\cos(n\pi x))^2 \, dx = \frac{\frac{1}{2}\cos(n\pi x)\sin(n\pi x) + \frac{1}{2}n\pi x}{n\pi} \Big|_{-1}^{1} = 1$$

Also (p_0, p_0) equals

$$\int_{-1}^{1} 1 \, dx = 2$$

(c) $f_o(x) = \frac{1}{3} + \sum_{k=1}^{n} [\frac{4}{k^2\pi^2}](-1)^k \cos k\pi x.$

6.45 **(a)**

$$\int_{-1}^{1} \cos(n\pi x) \sin(m\pi x)dx$$

$$= \left(-\frac{1}{2} \frac{\cos((n\pi + m\pi)x)}{n\pi + m\pi} + \frac{1}{2} \frac{\cos((n\pi - m\pi)x)}{n\pi - m\pi} \right) \Big|_{-1}^{1} = 0$$

(b) Since $h(x) = (f(x) + g(x))/2$ where $f(x)$ is the rasp from the text and $g(x)$ is the saw tooth function we see

$$h(x) = \frac{1}{4} + \sum_{k=1}^{n} \frac{((-1)^k - 1)}{k^2\pi^2} \cos k\pi x$$

$$+ \sum_{k=1}^{n} \frac{(-1)^{k+1}}{k\pi} \sin k\pi x$$

6.47 $c_k = ((-1)^{k+1}(0.002))/(k\pi).$

6.48 **(a)** 18. **(b)** You must prove that properties (a)–(d) from Theorem 6.12 on page 336 hold, where f, g, and h now represent elements of P_2. For the most part these are simple. For example, to prove property (b) we reason

$$(f+g,h) = (f(0)+g(0))h(0) + (f(1)+g(1))h(1) + (f(2)+g(2))h(2)$$
$$= (f(0)h(0) + g(0)h(0)) + (f(1)h(1) + g(1)h(1))$$
$$+ (f(2)h(2) + g(2)h(2))$$
$$= f(0)h(0) + f(1)h(1) + f(2)h(2) + g(0)h(0)) + g(1)h(1)$$
$$+ g(2)h(2)$$
$$= (f,h) + (g,h)$$

For property (d):

$$(f,f) = f(0)^2 + f(1)^2 + f(2)^2 \geq 0.$$

Furthermore, if $(f,f) = 0$ then $f(0) = f(1) = f(2) = 0$. A degree 2 polynomial $f(x)$ that is zero at three different values of x must be the zero polynomial. (d) $c_1 = 0, c_2 = -1, c_3 = 2$.

6.54 (b) The proof is as follows:

$$\|V+W\|^2 + \|V-W\|^2 = (V+W, V+W) + (V-W, V-W)$$
$$= (\|V\|^2 + 2(V,W) + \|W\|^2)$$
$$+ (\|V\|^2 - 2(V,W) + \|W\|^2)$$
$$= 2\|V\|^2 + 2\|W\|^2$$

6.56 From orthogonality

$$(W, cV_i) = (c_1 V_1 + \cdots + c_n, V_n, c_i V_i)$$
$$= c_i^2(V_i, V_i) = c_i^2 |V_i|^2$$

Hence

$$(W, W) = (W, c_1 V_1) + \cdots + (W, c_n V_n)$$
$$= c_1^2 |V_1|^2 + \cdots + c_n^2 |V_n|^2$$

proving the assertion.

6.4 ORTHOGONAL MATRICES

Problems begin on page **364**

EXERCISES

6.68 $|AX| = |[7, \sqrt{2}, 3]^t| = \sqrt{60} = |X|$

6.69 Compute $(R_\theta^x)^t R_\theta^x$, $(R_\theta^y)^t R_\theta^y$, and $(R_\theta^z)^t R_\theta^z$ to see that you get I.

6.71 (a) It is easily see that the columns of A mutually are orthogonal with length 17. Hence $c = \pm\frac{1}{17}$.

6.74 $|(AB)X| = |A(BX)| = |BX| = |X|$.

6.75 $(AB)(AB)^t = ABB^tA^t = AA^t = I$.

6.83

(b) $\frac{1}{11}\begin{bmatrix} 9 & -2 & 6 \\ -2 & 9 & 6 \\ 6 & 6 & -7 \end{bmatrix}$

6.84 Note that P is only unique up to scalar multiples. Since M_P is orthogonal, $k = |X|$. From figure 6.21 on page 362 we may take $P = X - X' = X - |X|I_1$. (See Exercise 6.85.) Hence the answers are

(a) $[3, 4]^t - [5, 0]^t = [-2, 4]^t$.

6.86 (a)

$$M_P = \begin{bmatrix} \frac{\sqrt{2}}{2} & 0 & \frac{\sqrt{2}}{2} \\ 0 & 1 & 0 \\ \frac{\sqrt{2}}{2} & 0 & -1/2\sqrt{2} \end{bmatrix}$$

$$M_P A = \begin{bmatrix} \sqrt{2} & \frac{\sqrt{2}}{2} & 2\sqrt{2} \\ 0 & -2 & 1 \\ 0 & \frac{\sqrt{2}}{2} & -\sqrt{2} \end{bmatrix}$$

(b)

$$M_Q = \begin{bmatrix} -2/3\sqrt{2} & 1/3 \\ 1/3 & 2/3\sqrt{2} \end{bmatrix}, \quad \begin{bmatrix} u & v \\ 0 & w \end{bmatrix} = \begin{bmatrix} 3/2\sqrt{2} & -\sqrt{2} \\ 0 & -1 \end{bmatrix}$$

6.5 LEAST SQUARES

Problems begin on page **377**

EXERCISES

6.89 The normal equation is $A^tAX = A^tB$. $B - B_0$ will be perpendicular to the columns of A if and only if $A^t(B - B_0) = 0$. We find $X = [30.9306, 2.0463]^t$, $B_0 = [31.9537, 33.1815, 34.0, 35.2278, 35.637]^t$.

6.91 (a)

$$A = \begin{bmatrix} 1 & 0.5 & 0.25 \\ 1 & 1.1 & 1.21 \\ 1 & 1.5 & 2.25 \\ 1 & 2.1 & 4.41 \\ 1 & 2.3 & 5.29 \end{bmatrix}, \quad B = \begin{bmatrix} 32.0, 33.0, 34.2, 35.1, 35.7 \end{bmatrix}$$

(b) $[a, b, c]^t = [30.9622, 1.98957, 0.019944]^t$.

$$A^tA = \begin{bmatrix} 5.0000 & 7.5000 & 13.4100 \\ 7.5000 & 13.4100 & 26.2590 \\ 13.4100 & 26.2590 & 54.0213 \end{bmatrix} \quad A^tB = \begin{bmatrix} 170.0000 \\ 259.4200 \\ 468.5240 \end{bmatrix}$$

6.93

$$A = \begin{bmatrix} 1 & 0 \\ 1 & 5 \\ 1 & 10 \\ 1 & 15 \end{bmatrix} \quad B = \begin{bmatrix} 5.4250 \\ 5.4681 \\ 5.5175 \\ 5.5683 \end{bmatrix}$$

Then

$$A'A = \begin{bmatrix} 4 & 30 \\ 30 & 350 \end{bmatrix} \qquad A'B = \begin{bmatrix} 21.9788 \\ 166.0400 \end{bmatrix}$$

Producing $[\ln a, b]' = [5.4228, 0.0096]'$. Then $P = 226.512e^{.00956t}$. In 2010, $P = 301.750$.

6.95 $B = [1, 2, 3, 4, 5]'$. The projection matrix, $\text{Proj}_W B$ and $\text{Orth}_W B$ are, respectively,

$$P = \frac{1}{14} \begin{bmatrix} 3 & 4 & 2 & 3 & 2 \\ 4 & 10 & -2 & 4 & -2 \\ 2 & -2 & 6 & 2 & 6 \\ 3 & 4 & 2 & 3 & 2 \\ 2 & -2 & 6 & 2 & 6 \end{bmatrix}, \qquad \frac{1}{14} \begin{bmatrix} -25 \\ 4 \\ -12 \\ 17 \\ 16 \end{bmatrix}, \qquad \frac{1}{14} \begin{bmatrix} 39 \\ 24 \\ 54 \\ 39 \\ 54 \end{bmatrix}$$

6.100 A is not be invertible.

6.6 QUADRATIC FORMS: ORTHOGONAL DIAGONALIZATION

Problems begin on page **392**

EXERCISES

6.105 The only matrices that describe $X'AX = d$ (with the same d) are as follows. If d is allowed to change, then any non-zero scalar multiple of this matrix also works.

$$\begin{bmatrix} 3 & a \\ 2-a & 3 \end{bmatrix}$$

6.106 $D = \begin{bmatrix} 18 & 0 & 0 \\ 0 & 18 & 0 \\ 0 & 0 & 9 \end{bmatrix}, \qquad Q = \frac{1}{30} \begin{bmatrix} -4\sqrt{5} & -4\sqrt{5} & 10\sqrt{6} \\ 10\sqrt{5} & -8\sqrt{5} & 5\sqrt{6} \\ -8\sqrt{5} & 10\sqrt{5} & 5\sqrt{6} \end{bmatrix}$

6.107 We give the standard forms followed by the corresponding orthogonal matrix Q. The columns are the bases.

(a) $-4(x')^2 + 6(y')^2 = 1$

$$Q = \frac{1}{\sqrt{5}} \begin{bmatrix} 1 & 2 \\ -2 & 1 \end{bmatrix}$$

(b) $2(x_2')^2 + 4(x_3')^2 = 4$

$$Q = \frac{1}{2} \begin{bmatrix} 1 & \sqrt{2} & 1 \\ -\sqrt{2} & 0 & \sqrt{2} \\ 1 & -\sqrt{2} & 1 \end{bmatrix}$$

6.108 For this exercise one needs the eigenvalues but not the eigenvectors.
(a) Ellipse.

$$\left(3 + \sqrt{2}\right)x^2 + \left(3 - \sqrt{2}\right)y^2 = 1$$

(c) A pair of lines.

$$10\,y^2 = 1$$

(d) Ellipse.

$$10x^2 + 15y^2 = 1$$

6.109 The basis is the columns of the given matrices Q.
(a)

$$Q = \frac{1}{\sqrt{5}} \begin{bmatrix} -1 & 2 \\ 2 & 1 \end{bmatrix}$$

6.111 The answers are (a) a hyperbola, (b) the empty set, and (c) a pair of lines.

In general for a 2×2, non-zero, symmetric matrix A, the graph of $X^t A X = d$ for $d > 0$ is an ellipse if both eigenvalues are positive, the empty set if both are negative, and a pair of lines if one of the eigenvalues is 0.

6.112 $\frac{118}{25}x^2 + \frac{48}{25}xy + \frac{132}{25}y^2 = 1$. In rotated coordinates our ellipse is given by

$$\frac{(x')^2}{9} + \frac{(y')^2}{4} = 1 \Leftrightarrow 4(x')^2 + 9(y')^2 = 36$$

We let

$$Q = \frac{1}{5}\begin{bmatrix} 3 & -4 \\ 4 & 3 \end{bmatrix}, \quad D = \begin{bmatrix} 4 & 0 \\ 0 & 9 \end{bmatrix}$$

Since

$$QDQ^t = \frac{1}{5}\begin{bmatrix} 36 & -12 \\ -12 & 29 \end{bmatrix}$$

the equation

$$36x^2 - 24xy + 29y^2 = 5 \cdot 36 = 180$$

gives the desired ellipse.

6.113 Let Q be as in Exercise (6.112) and D be as below where $c, d > 0$. Then the matrix of the quadratic form is

$$D = \begin{bmatrix} c & 0 \\ 0 & -d \end{bmatrix}, \quad QDQ^t = \begin{bmatrix} 9c - 16d & 12c + 12d \\ 12c + 12d & 16c - 9d \end{bmatrix}$$

6.116 This is a direct consequence of (6.81) on page 386 and the expression for D given in this example.

6.118 From the solution to Exercise 6.114, any matrix $\begin{bmatrix} a & b \\ b & c \end{bmatrix}$ with positive entries such that $ac < b^2$ will work.

6.120

$$X^tAX = X^tB^tBX = (BX) \cdot (BX) = |BX|^2 \geq 0.$$

Assume that A is positive semidefinite and X is an eigenvector for A with eigenvalue λ. Then

$$0 \le X^t A X = X^t(\lambda X) = \lambda|X|^2$$

Thus the eigenvalues of A are non-negative.

6.122

$$A^t = (QDQ^{-1})^t = (QDQ^t)^t$$
$$= (Q^t)^t D^t Q^t = QDQ^{-1} = A$$

6.7 THE SINGULAR VALUE DECOMPOSITION (SVD)

Problems begin on page **404**

EXERCISES

6.126 **(a)**

$$U = \frac{1}{\sqrt{2}}\begin{bmatrix}-1\\1\end{bmatrix}, \quad E = [2]$$

(c)

$$U = \frac{1}{\sqrt{15}}\begin{bmatrix}-2\\-1\\-3\\1\end{bmatrix}, \quad E = [15]$$

6.127 **(a)**

$$U = \frac{1}{\sqrt{2}}\begin{bmatrix}1 & -1\\1 & 1\end{bmatrix}, \quad D = \begin{bmatrix}4 & 0\\0 & \sqrt{6}\end{bmatrix}$$

$$V = \begin{bmatrix}\frac{\sqrt{2}}{2} & \frac{\sqrt{3}}{3}\\0 & \frac{\sqrt{3}}{3}\\\frac{\sqrt{2}}{2} & \frac{\sqrt{3}}{3}\end{bmatrix}$$

(b)

$$U = \frac{1}{\sqrt{5}} \begin{bmatrix} 2 & 1 \\ 1 & -2 \end{bmatrix}, \quad D = \begin{bmatrix} 2\sqrt{15} & 0 \\ 0 & \sqrt{15} \end{bmatrix}$$

$$V = \frac{1}{\sqrt{3}} \begin{bmatrix} -2 & 2 \\ 2 & 1 \\ 1 & 2 \end{bmatrix}$$

(c)

$$U = \frac{1}{3} \begin{bmatrix} 2 & 1 \\ 1 & 2 \\ 2 & -2 \end{bmatrix}, \quad D = \begin{bmatrix} 2 & 0 \\ 0 & 1 \end{bmatrix}$$

$$V = \frac{1}{5} \begin{bmatrix} 4 & -3 \\ 3 & 4 \end{bmatrix}$$

6.130 (a) $[0, 1, 1]^t$, (c) $\left[\frac{29}{9}, \frac{71}{27}\right]^t$.

6.8 HERMITIAN SYMMETRIC AND UNITARY MATRICES

Problems begin on on page **417**

EXERCISES

6.140 (a) $3 + 20i$, (c) $4 + 6i$.

6.141 **(b)**

$$B^*A^* = (\overline{B})^t(\overline{A})^t = (\overline{A}\,\overline{B})^t = (\overline{AB})^t = (AB)^*$$

6.142 **(a)**

$$\begin{bmatrix} 3 & 2 \\ -2 & 3 \end{bmatrix}$$

6.144 **(a)** From Theorem 6.32 on page 414 the eigenvalues λ and β are real. From formula (6.115) on page 413

$$\lambda H(Z, W) = H(\lambda Z, W) = H(AZ, W)$$
$$= H(Z, A^*W) = H(Z, AW) = H(Z, \beta W)$$
$$= \bar{\beta} H(Z, W) = \beta H(Z, W)$$

Hence $(\lambda - \beta)H(Z, W) = 0$ which implies $H(Z, W) = 0$.

6.150 **(a)**

$$(1/\sqrt{6}) \begin{bmatrix} 1+i & \sqrt{2}(1+i) \\ -2 & \sqrt{2} \end{bmatrix}, \ D = \begin{bmatrix} 0 & 0 \\ 0 & 3 \end{bmatrix}$$

6.151 $$(cA)^* = [\overline{ca_{ji}}] = [\bar{c}\,\overline{a_{ji}}] = \bar{c}A^*$$

6.152 Let X be an eigenvector for U. Then from 6.116 on page 413

$$< X, X > = < UX, UX >$$
$$= < \lambda X, \lambda X >$$
$$= \lambda \bar{\lambda} < X, X >$$
$$= |\lambda|^2 < X, X >$$

Since $X \neq 0$, $< X, X > \neq 0$, so we can divide both sides of the above equality by $< X, X >$, proving that $|\lambda|^2 = 1$; hence $|\lambda| = 1$.

6.155 **(a)** Let $W = AZ$. Then $0 = H(AZ, AZ) = |AZ|^2$ Hence $AZ = 0$ for all Z implying that $A = 0$.

6.158 The proof is the same as that for Proposition 6.4 on page 363 with transposes replaced by adjoints.

CHAPTER 7

GENERALIZED EIGENVECTORS

7.1 GENERALIZED EIGENVECTORS

Problems begin on page **429**

EXERCISES

7.1 In our answers, for any sequence of vectors Y_i, $(A - \lambda I)Y_i = Y_{i-1}$ where for $i < 0$, $Y_i = 0$. Some of these answers differ from the answers at the back of the text. This is done deliberately to stress that the chain bases are not unique. For example adding an eigenvector corresponding to the given eigenvalue to any chain basis element produces another chain basis, provided the result is not the zero vector. We give the relation between these answers and the answers X, Y, Z, and W from the back of the text.

(**a**) $\qquad \lambda = 3, X_0 = \begin{bmatrix} 1 \\ 0 \\ 0 \end{bmatrix}, X_1 = \begin{bmatrix} 0 \\ 1 \\ 1 \end{bmatrix}, \ \lambda = 2, \ W_0 = \begin{bmatrix} 1 \\ -1 \\ 0 \end{bmatrix}$

$$Y = X_0, \ X = -W_0, \ Z = X_1 - X_0$$

Solutions Manual to Accompany Linear Algebra: Ideas and Applications, Fourth Edition. Richard Penney.
© 2016 John Wiley & Sons, Inc. Published 2016 by John Wiley & Sons, Inc.

(e)

$$\lambda = 2, \, X_0 = \begin{bmatrix} 1 \\ 0 \\ 0 \\ 0 \end{bmatrix}, \, X_1 = \begin{bmatrix} 0 \\ -1 \\ -1 \\ 0 \end{bmatrix},$$

$$U_0 = \begin{bmatrix} 0 \\ 0 \\ -1 \\ 1 \end{bmatrix}, \, \lambda = 1, \, W_0 = \begin{bmatrix} 2 \\ 1 \\ 0 \\ 0 \end{bmatrix}$$

$$X = -\frac{1}{2}W_0, \, Y = -X_0, \, Z = -X_1 + U_0 + X_0$$

(g)

$$\lambda = 1, \, X_0 = \begin{bmatrix} 1 \\ 0 \\ 0 \\ 0 \end{bmatrix}, \, X_1 = \begin{bmatrix} 0 \\ \frac{5}{9} \\ -\frac{2}{9} \\ \frac{1}{9} \end{bmatrix}$$

$$\lambda = 2, \, W_0 = \begin{bmatrix} 2 \\ 1 \\ 0 \\ 0 \end{bmatrix}, \, W_1 = \begin{bmatrix} -\frac{5}{3} \\ 0 \\ \frac{1}{3} \\ 0 \end{bmatrix}$$

$$X = X_0, \, Y = 18X_0 + 9X_1, \, Z = \frac{1}{2}W_0, \, W = \frac{7}{4}W_0 - \frac{3}{2}W_1$$

7.2 Let notation be as in the solution to Exercise 7.1.

(a) $B = X_0 + X_1$. From (7.7) on page 423

$$A^n B = A^n X_0 + A^n X_1$$
$$= 3^n X_0 + 3^n X_1 + n3^{n-1}X_0$$
$$= (3^n + n3^{n-1})X_0 + 3^n X_1$$

(b) $B = X_0 + 2X_1 + X_2$. From (7.7) on page 423

$$
\begin{aligned}
A^n B &= A^n X_0 + 2A^n X_1 + A^n X_2 \\
&= 3^n X_0 + 2(3^n X_1 + n3^{n-1} X_0) \\
&\quad + (3^n X_2 + n3^{n-1} X_2 + \frac{n(n-1)}{2} 3^{n-2} X_0) \\
&= \left(2n3^{n-1} + 3^n + n(n-1)3^{n-2}/2\right) X_0 \\
&\quad + (n3^{n-1} + 2\,3^n)X_1 + 3^n X_2
\end{aligned}
$$

(c) $B = X_0 + X_1$

$$
\begin{aligned}
A^n B &= A^n X_0 + A^n X_1 \\
&= 3^n X_0 + 3^n X_1 + n3^{n-1} X_0 \\
&= (3^n + n3^{n-1})X_0 + 3^n X_1
\end{aligned}
$$

(d) $B = -W - 2X_0 - 9X_1$

$$
\begin{aligned}
A^n B &= -A^n W - 2A^n X_0 - 9A^n X_1 \\
&= -3^n W - 2^{n+1} X_0 - 2^n 9X_1 - n2^{n-1}9X_0 \\
&= -3^n W - (2^{n+1} + 9n2^{n-1})X_0 - 2^n 9X_1
\end{aligned}
$$

(e) $B = -W_0 + 3X_0 - 2X_1 + U_0$

$$
\begin{aligned}
A^n B &= -A^n W_0 + 3A^n X_0 - 2A^n X_1 + A^n U_0 \\
&= -W_0 + 2^n 3X_0 - 2^{n+1} X_1 - n2^n X_0 + 2^n U_0 \\
&= -W_0 + (3-n)2^n X_0 - 2^{n+1} X_1 + 2^n U_0
\end{aligned}
$$

(f) $B = 3W_0 + X_0 + 3X_1 + 2X_2$

$$
\begin{aligned}
A^n B &= A^n W_0 + A^n X_0 + 3A^n X_1 + 2A^n X_2 \\
&= 2^n W_0 + X_0 + 3(X_1 + nX_0) + 2\left(X_2 + nX_1 + \frac{n(n-1)}{2}X_0\right) \\
&= 2^n W_0 + (1 + 3n + n(n-1))X_0 + (3+2n)X_1 + 2X_2
\end{aligned}
$$

(g) $B = 24X_0 + 9X_1 + 3W_0 - 4W_1$

$$A^n B = 24A^n X_0 + 9A^n X_1 + 3A^n W_0 - 4A^n W_1$$
$$= 24X_0 + 9(X_1 + nX_0) + 2^n 3W_0 - 4\left(2^n W_1 + n2^{n-1} W_0\right)$$
$$= (24 + n)X_0 + 9X_1 + (3\,2^n - 4n2^{n-1})W_0 - 2^{n+2} W_1$$

(h) $B = \frac{1}{9} W_0 - \frac{1}{18} X_0 + \frac{1}{3} X_1$

$$A^n B = \frac{1}{9} A^n W_0 - \frac{1}{18} A^n X_0 + \frac{1}{3} A^n X_1$$
$$= \frac{2^n}{9} W_0 - \frac{(-1)^n}{18} X_0 + \frac{1}{3}\left((-1)^n X_1 + n(-1)^{n-1} X_0\right)$$
$$= \frac{2^n}{9} W_0 + \left(\frac{(-1)^{n-1} n}{3} - \frac{(-1)^n}{18}\right) X_0 + \frac{(-1)^n}{3} X_1$$

7.4 Let $B = A - 4I$. Then

(b) Write B and W as partitioned matrices

$$B = \begin{bmatrix} C & D \\ 0 & E \end{bmatrix}, E = \begin{bmatrix} 1 & j \\ 0 & 1 \end{bmatrix}, W = \begin{bmatrix} W_0 \\ U \end{bmatrix}$$

where C is 3×3, D is 3×2, 0 is 2×3, $W_0 = [x, y, z]'$ and $U = [u, v]'$. Then for $n \in \mathbb{N}$ and some $W_n \in \mathbb{R}^3$

$$B^n W = \begin{bmatrix} W_n \\ E^n U \end{bmatrix}$$

Since E is invertible, $B^n W = 0$ for some n implies $U = 0$ as claimed.

7.6 The vector $X_0 = [1, 0, 0]'$ is an eigenvector with eigenvalue 5. The general solution to $(A - 5I)X = \pm X_0$ is $X = \pm[t, 1, 0]'$. In particular $X_1 = [0, 1, 0]'$ and $X_2 = [1, -1, 0]'$ are both order two generalized eigenvectors and $X_1 + X_2 = X_0$.

7.9 Assume first that X is a λ-eigenvector for A. Then for some k

$$0 = (A - \beta I)^k X = (\lambda - \beta)^k X$$

which implies that $\beta = \lambda$.

Thus we may assume that X is an order k generalized eigenvector for A where $k > 1$. Then $X_o = (A - \lambda I)^{k-1}X$ is a λ-eigenvector. From Exercise 7.8 with $B = (A - \beta I)^{k-1}$, X_o is also a generalized eigenvector for eigenvalue β. Hence $\beta = \lambda$,

(a) Note that

$$P^{-1}(A - \lambda I)P = P^{-1}AP - \lambda P^{-1}IP = B - \lambda I$$

Hence $P^{-1}X \in \mathbb{K}^n(B, \lambda)$ if and only if there is a k such that

$$0 = (B - \lambda I)^k P^{-1}X = P^{-1}(A - \lambda I)^k P(P^{-1}X) = P^{-1}(A - \lambda I)^k X$$

This is zero if and only if $(A - \lambda I)^k X = 0$, proving the claim.

7.11 (a)

$$\begin{aligned}
(B + C)^3 &= (B + C)(B + C)^2 = B(B + C)^2 + C(B + C)^2 \\
&= B(B^2 + 2BC + C^2) + C(B^2 + 2BC + C^2) \\
&= B^3 + 2B^2C + BC^2 + CB^2 + 2CBC + C^3 \\
&= B^3 + 3B^2C + 3BC^2 + C^3
\end{aligned}$$

7.2 CHAIN BASES

Problems begin on page **443**

EXERCISES

7.12 (a)

$$J = \begin{bmatrix} 3 & 1 & 0 \\ 0 & 3 & 0 \\ 0 & 0 & 2 \end{bmatrix}, \quad Q = \begin{bmatrix} 1 & 0 & 1 \\ 0 & 1 & -1 \\ 0 & 1 & 0 \end{bmatrix}$$

(e)

$$J = \begin{bmatrix} 2 & 1 & 0 & 0 \\ 0 & 2 & 0 & 0 \\ 0 & 0 & 2 & 0 \\ 0 & 0 & 0 & 1 \end{bmatrix}, \quad Q = \begin{bmatrix} 1 & 0 & 0 & 2 \\ 0 & -1 & 0 & 1 \\ 0 & -1 & -1 & 0 \\ 0 & 0 & 1 & 0 \end{bmatrix}$$

(g)

$$Q = \begin{bmatrix} 1 & 1 & 0 & 0 \\ 0 & 1 & 0 & 0 \\ 0 & 0 & 2 & 1 \\ 0 & 0 & 0 & 2 \end{bmatrix}, \quad J = \begin{bmatrix} 1 & 0 & 2 & -5/3 \\ 0 & 5/9 & 1 & 0 \\ 0 & -2/9 & 0 & 1/3 \\ 0 & 1/9 & 0 & 0 \end{bmatrix}$$

7.14 (a)

$$\begin{bmatrix} 3 & 1 & 0 & 0 & 0 & 0 & 0 & 0 \\ 0 & 3 & 1 & 0 & 0 & 0 & 0 & 0 \\ 0 & 0 & 3 & 0 & 0 & 0 & 0 & 0 \\ 0 & 0 & 0 & 4 & 0 & 0 & 0 & 0 \\ 0 & 0 & 0 & 0 & 6 & 0 & 0 & 0 \\ 0 & 0 & 0 & 0 & 0 & 6 & 1 & 0 \\ 0 & 0 & 0 & 0 & 0 & 0 & 6 & 1 \\ 0 & 0 & 0 & 0 & 0 & 0 & 0 & 6 \end{bmatrix}$$

7.18 Assume that the result fails. Let m be smallest natural number for which there exist $X_i \in \mathbb{K}^n(A, \lambda_i)$, $1 \le i \le m$, not all equal to zero, such that

$$X_1 + \cdots + X_m = 0.$$

In this case, none of the X_i are zero since we may obtain a smaller value of m by omitting any one of the X_i which equals 0. Let k be the order of X_m. Then

$$\begin{aligned} 0 &= (A - \lambda_m I)^k X_1 + \cdots + (A - \lambda_m I)^k X_m \\ &= (A - \lambda_m I)^k X_1 + \cdots + (A - \lambda_{m-1} I)^k X_{m-1} \end{aligned}$$

From Exercise 7.8 on page 430 with $B = (A - \lambda_m I)^k$ the vectors $Y_i = (A - \lambda_m I)^k X_i$ belong to $\mathbb{K}^n(A, \lambda_i)$. Hence from the choice of m, $Y_i = 0$ for all i. Then Exercise 7.9 on page 430 shows that $X_i = 0$ for all i, contradicting the assumption that the result is false.

7.19 From Theorem 5.1 on page 275, $n_1 + \cdots + n_k = n$. Hence it suffices to show that $\mathcal{B}$ is an independent set. Thus suppose that c^i_j, $1 \le j \le n_i$ are scalars such that

$$\sum_{i=1}^{m} \sum_{j=1}^{n_i} c^i_j Y^i_j = 0$$

where $\mathcal{B}_i = \{Y^i_1, \ldots, Y^i_{n_i}\}$. It follows from Exercise 7.18 with

$$X_i = \sum_{j=1}^{n_i} c^i_j Y^i_j$$

that $X_i = 0$ for all i. Hence, from the independence of $\mathcal{B}_i$, $c^i_j = 0$ for all i and j proving the result.

7.20 (e) Note that as a partitioned matrix

$$C = \begin{bmatrix} E & 0 \\ 0 & F \end{bmatrix}$$

where $E = 2I + A$ and $F = -I + B$ with A and B as given at the beginning of the problem. Then

$$C^5 = \begin{bmatrix} E^5 & 0 \\ 0 & F^5 \end{bmatrix}$$

where E^5 is given in part (b) with $\lambda = 2$ and F^5 is given in part (d) with $\lambda = -1$.

CHAPTER 8

NUMERICAL TECHNIQUES

8.1 CONDITION NUMBER

Problems begin on page **451**

8.1 For $|X|$ and $|Y|$, see the answers to Section 6.1, Exercise 6.1, on page 85 of this manual.

    (a) $|X|_\infty = \max\{3, 4\} = 4$, $|Y|_\infty = \max\{|-1|, 2\} = 2$, $|X|_1 = 3 + 4 = 7$, $|Y|_1 = |-1| + 2 = 3$.

    (c) Same as (a).

    (e) $|X|_\infty = 4$, $|Y|_\infty = 8$, $|X|_1 = 7$, $|Y|_1 = 14$.

    (g) $|X|_\infty = 6$, $|Y|_\infty = 2$, $|X|_1 = 16$, $|Y|_1 = 9$.

8.3 **(c)** From Theorem 6.3 on page 311 with $Y = [1, 1, \ldots, 1]^t$ and $X_o = [|x_1|, \ldots, |x_n|]^t$

$$|X|_1 = Y \cdot X_o \le |Y|\,|X_o| = \sqrt{n}\,|X|$$

Solutions Manual to Accompany Linear Algebra: Ideas and Applications, Fourth Edition. Richard Penney.
© 2016 John Wiley & Sons, Inc. Published 2016 by John Wiley & Sons, Inc.

8.4 By definition if $A = [A_1, A_2, \ldots, A_n]$ where A_i are columns, then $\|A\| = |B|_\infty$ where $B = [|A_1|_1, \ldots, |A_n|_1]$. In each part, we give B followed by $\|A\| = \|B\|_\infty$.

   (a) $[7, 9, 9]$, 9, (c) $[6, 12, 18]$, 18, (e) $[13, 24, 22, 23]$, 24, (g) $[5, 12, 2, 10]$, 12.

8.5 (c) and (e) are not invertible. In each part we write the answer as $\|A\| \, \|A^{-1}\|$.

   (a) $9 \left(\frac{29}{5} \right) = 52.20$, (d) $4 \cdot 2 = 8$, (g) $12 \frac{49}{3} = 196$.

8.7 Let $A = [A_1, \ldots, A_n]$ and $A^{-1} = [C_1, \ldots, C_n]$ where the A_i and C_i are columns. Let j and k be such that $|A_j|_1 = \|A\|$ and $|C_k|_1 = \|A^{-1}\|$. Then formula (8.9) on page 450 is an equality $\Delta B = I_k$ and $\Delta X = C_k$ while formula (8.10) is an equality with $X = I_j$ and $B = A_j$. It follows that

$$\frac{|\Delta X|_1}{|X|_1} = \frac{\|A^{-1}\| \, |\Delta B|_1}{|B|_1 / \|A\|} = \|A^{-1}\| \, \|A\| \, \frac{|\Delta B|_1}{|B|_1}$$

8.9 (b) cond$A = 35,022$, $B = [77, 61, 213]^t$, $\Delta B = [0, .001, 0]^t$.

8.10 Let $A = [A_1, \ldots, A_n]$ be an orthogonal matrix where the A_i are columns. From Exercise 8.3.(c), $|A_i|_1 \le \sqrt{n}|A_i| = \sqrt{n}$. Since A^{-1} is also orthogonal, $\|A^{-1}\| \le \sqrt{n}$ as well. The result follows.

8.2 COMPUTING EIGENVALUES

Problems begin on page **462**

EXERCISES

8.11 Part (b): For each matrix in Exercise 5.3 on page 280 we give $C^6B \cdot C^5B / C^5B \cdot C^5B$ where $C = A^{-1}$. (For small dimensional matrices and small k there is no need to use the LU factorization.)

   (b) $[1024, 992, 992]^t / 32\sqrt{2946} = [0.5895, .5710, 0.5710]^t$, $12193792/3016704 = 4.042$.

8.13 **(a)** The eigenvalues are $4.1102, 0.0895, -2.1543, -5.0455$. Hence the dominant eigenvalue is -5.0455.

(b) $k = 35$ works.

(c) We use analogous notation to that used in Example 8.4 on page 459. The dominant eigenvalue is near -5. Hence we use $C = A + 6I$ in our computations. We find

$$C_5 - 6I = \begin{bmatrix} 3.3784 & 0.90369 & -14.0479 \\ -1.534 & 4.622 & 3.8360 \\ 0.000 & 0.000 & -4.0 \end{bmatrix}$$

8.14 **(a)** As partitioned matrices the general $n \times n$ matrix A and element X of $\mathbb{R}^n_i$ may be written respectively as

$$A = \begin{bmatrix} C_i & D_i \\ F_i & E_i \end{bmatrix}, \quad X = \begin{bmatrix} X_i \\ 0 \end{bmatrix}$$

where $X_i \in \mathbb{R}^i$ and C_i is an $i \times i$ matrix. Then

$$AX = \begin{bmatrix} C_i X_i \\ F_i X_i \end{bmatrix}$$

Hence AX belongs to $\mathbb{R}^n_i$ for all X_i if and only if $F_i = 0$. This is true for all i if and only if A is upper triangular.

(b) For $i = n$ the result is clear since $\mathbb{R}^n_i = \mathbb{R}^n$ for $i \geq n$. Thus we assume $i < n$. As partitioned matrices the general $n \times n$ matrix A and element X_i of $\mathbb{R}^n_i$ may be written respectively as

$$A = \begin{bmatrix} C & c \\ B & D \end{bmatrix}, \quad X = \begin{bmatrix} X_i \\ 0 \end{bmatrix}$$

where B is $(n-1) \times (n-1)$, C is $1 \times (n-1)$, D is $(n-1) \times 1$, $X_i \in \mathbb{R}^{n-1}_i$, and c is a scalar. Note that A is upper Hesssian if and only if B is upper triangular. Since

$$AX = \begin{bmatrix} CX_i \\ BX_i \end{bmatrix}$$

our result follows from part (a) along with the observation that $AX \in \mathbb{R}^n_{i+1}$ if and only if $BX_i \in \mathbb{R}^{n-1}_i$.

(c) This is immediate from parts (a) and (b) and the observations that for $X \in \mathbb{R}_i^n$ $AR(X) = A(R(X))$ and $RA(X) = R(A(X))$.

(d) This follows from part (c), the observation that the inverse of an upper triangular matrix is upper triangular, and the equalities $Q = AR^{-1}$ and $RQ = RAR^{-1}$.

(e) Immediate from part (d).